StarFleet Assembly Manuals

Ultimate Modeling Edition
by Paul M. Newitt

Other books from CultTVman

CultTVman's Ultimate Modeling Guide to Classic Sci-Fi Movies

CultTVman's Ultimate Modeling Guide to the Jupiter II

NCC-1701

NCC-1701

NCC-1706

NCC-1701

StarFleet Assembly Manuals

Original SFAM book Design by Paul M. Newitt with updated book design by Melanie McMahon.

Published by CultTVman, Lawrenceville, GA
www.culttvman.com
First Printing May 2004
ISBN 0-9701455-2-7

Library of Congress-in-Publication Data
Newitt, Paul M.
"StarFleet Assembly Manuals"
Library of Congress Control Number: 2004104823

Contents

Foreword

Imagine being able to build a model that actually looks like that famous starship on television!

Just over 25 years ago, Paul Newitt wrote the first of the StarFleet Assembly Manual series, the first publication aimed at science fiction modelers. It used photos, diagrams and detailed text to illustrate basic modeling techniques, teaching people like me how to accurately build starship models. No one had ever done this before! His book, and the follow-ups, changed the modeling hobby forever.

Today, the opportunity of taking a toy-like model kit and making it into a believable starship still exists. Thanks to the people at Polar Lights, we now have new versions of these kits to build! The information in this book can be applied to today's starship models, as well as any other SF you may work on. Although the original AMT Enterprise is long out of production, plenty of them are still sitting in closets, just waiting for glue and paint.

This volume collects the first three StarFleet Assembly Manuals pretty much as they were originally printed in the late 1970's. We've removed references to companies long since obsolete, and added some updated resources for today's model builders. You'll find additional information about the Polar Lights models as well as the original studio filming miniatures, too.

Now get out there and build a starship!

Steve Iverson
www.culttvman.com

Introduction

To the Trek model maker, enthusiast and collector,

It's been some 27 years since I wrote the first StarFleet Assembly Manual, and I am still awestruck by its continued popularity and interest. A new genre of scale model making was created when this book came out in 1977. Except for the special effects industry in Hollywood, no one had retrofitted any of the Star Trek model kits with electronics and fiber optic, and no one had presented those techniques and materials in book form. This was new territory, and has since spawned a whole new type of FX model making. SFAM ONE was the beginning, SFAM TWO introduced "battle damage", and SFAM THREE continued from the first book to create awesome starships in miniature. Accessory resin castings, electronics, and countless articles and web pages have arisen from these books during the past quarter century.

The SFAM series developed from my background in model railroading, and the techniques I had learned since I was about 13. Fiber optic was a new material, found in those exotic lamps, and LEDs were fairly new, too. In 1975, I edited two issues of a Sacramento club "fanzine" called "SubSpace." This publication was extremely popular at "The Federation Trading Post" stores in Berkeley, California, and New York. With the publishing expertise I'd learned, it was time to move on and do my own type of fanzine. I had built about ten (unlit) AMT Enterprise kits for local conventions, and sold many built up on wooden stands to fans who were not into model making, but wanted something for their home. A few fans suggested how neat it would be to install lights, and somehow make the rotating fan effect in the nacelles.

About this time, the Franz Joseph blueprints and later the StarFleet Technical Manual had come out, and was a strong influence to the look and feel of whatever book I decided to do. It was Franz Joseph's book design that inspired me to combine a book about building Trek models and make it look like a technical manual.

A friend and mentor of sorts in Sacramento, AJ Dezmeretz, talked me into doing these books in digest size, since his printing machines only did single-sheet size. It was his direction and support that helped me with my paste-ups and layouts, as well as how to shoot photos for halftones, and other details. It was from a S.T.A.R. Sacramento newsletter that I noticed the art of Al Sirois, of New Haven, Connecticutt., and asked him if he could do some cover and center section artwork for my new books. At that time, it was a stretch for him, but he pulled it off. His visions were right on the mark, predating any concept art for Star Trek: The Motion Picture in 1979. Al did the artwork for all three SFAM books, and his artwork captured the spirit of in-space assembly and battle damage with alien ships.

The first book sold fairly well, with a run of only a couple hundred copies and even piqued the interest of other Sci-Fi and Trek “dealers” on the east coast. This put my little book on the map, and the demand for new and more copies went from there.

SFAM TWO became a work-learn project for my Design major at U.C. Davis, and I received guidance from a wonderful graphic artist who helped me focus on better layout design that I use to this day. This project was completed on time and was published in late 1977. I wanted to make this version look different, as if it was information on the assembly of alien ships captured by StarFleet Intelligence. This was a time when the original series was still in endless reruns, when fanzines and fan art presented a lot of original thinking and wild speculation of what Trek could—or should—be. To know how to build a scale model from the Trek genre, and to light it up and animate it added to this grass-roots fandom that has fueled the interest all these years.

SFAM THREE was a big effort and it required a lot of rethinking about the AMT kit. Making ten LEDs flash in sequence, then breaking that up into even and odd groups of five for each nacelle was something I conjured up, and was exclusive to this book. The navigation lights were just simple flashers, which worked out well. The use of fiber optic was something new to me, but since the FX folks in Hollywood used literally miles of it, why not use it in an AMT kit? My first prototype was built for my (then) girlfriend. I threaded fiber optic into the saucer and hull, and lensed the ends. It all ran down a permanent brass support tube to the base and to a single light source. It was tightly packed, and crude by today's standards, but I remember thinking how this will become a big thing to do! The second prototype was built with an elaborate base using an inverted aquarium light to shine up on it, and it won an award at an Equicon convention in Los Angeles in 1978.

At that time, Edmund Scientific sold it in rolls of fiber optic for educational use, and .040" was perfect to simulate lighted ports. The challenge was getting the ends of the fiber optic to come together and finding a light source that would be bright enough. This, too was braving new territory in Trek model making, and I was able to find the best way to manipulate the optic to light the model.

I gave my list of circuit requirements to a (then) IBM technician from Silicon Valley, Terry Whittier, and the LS-100a circuit board was born. He devised this board to have the corners cut off to better fit into the saucer, and to support my request for a center rod to illuminate the sensor domes on the top and bottom. This circuit board worked well, and it was a successful project. Mimicked today by other enthusiasts for sale, it has proven its viability.

Since the release of all three SFAM books, many things have happened. In 1978, I contacted the west coast representative for AMT (then owned by Matchbox), to present one of my built up and lit starships to them. This meeting took place in a board room high in a skyscraper on prestigious Wilshire Boulevard in Los Angeles. The representative was immediately impressed with the starship and its lights. I was asked if there was a way to make those nacelle lights as separate units to mass market. Needless to say, I can't help but think that my presentation and books provided some sort of inspiration for the development of later AMT Trek kits with lights and sound.

Most of the many, many letters and calls I had

received were appreciative of how I was able to come up with a unique idea, and then explain it in easy to understand way. There were people from all backgrounds and ages. I can't say how many times I'd gotten hand drawn plans of how to achieve the rotating fan effect in the nacelles—flexible cables, detailed cutout domes, flashing lights in tiers, on and on and on. To this day, this electro-mechanical-modeling quandary is still unanswered to some. Other interesting mail came from other countries and places where I'd never think there was an interest. Even a few students of science that are now at NASA were part of my list of inquirers and customers. One east coast customer had his interest piqued by my books and later did the diagrams for superdetailing the exterior of the “E”. Little did I know at the time that he would become the premier science fiction model maker, with the highest professional standards; his name was David Merriman.

Many doors were opened by these books, including interest from New Yorkers Geoffrey Mandel and Doug Drexler. Geoff invited me to New York City for a Trek convention in the late summer of 1978, and at 22, I made the trip to the big apple to sell my SFAM books. A WHOLE briefcase of books that about broke the back of the cab driver that picked it up at the airport. This was also my introduction the the NY subway system—at 3am in Brooklyn! My friendship with Geoff and Doug continues today. Geoff is an author and freelance graphic designer for shows like JAG and NCIS, and Doug is an academy award winner for his makeup on Dick Tracy, and is the primary designer of the NX-01

"Enterprise". It's hard to measure the entire amount of influence that these books have had on fans and modelers.

SFAM ONE was designed for the fan that was NOT a modeler, but wanted a nice, presentable model in his or her room. What I DO know is that this was just the beginning of a whole new genre in model making, and it fueled other aspects of Sci-Fi modeling as well. What is important is how these books can take you beyond the dimension of a static model, and how it can bring life into it, and take you just a little bit closer to that "reality" of Trek. As my close friend Frank Kurson says, "because if you don't hurry, Star Trek will never really happen".

Unfortunately, over the years, these books found their way to being "pirated" and illegally copied (no fanzine was immune to unethical copying then). This special reprint edition is the best reproduction of these books ever offered, and far surpasses any previous version—original or copied. This is sure to become a true collectible. The features Steve and Jay have added bring new life to an already popular book series. It is with great excitement to reintroduce you to these wonderful books, and with Steve Iverson, to include new material to allow you to create your own special effects with your Trek model kits.

When you have the lights out, and your ship is ablaze in FX lighting, you will know it was worth the effort. The moment will speak for itself.

Thank you for all your support,
Paul M. Newitt

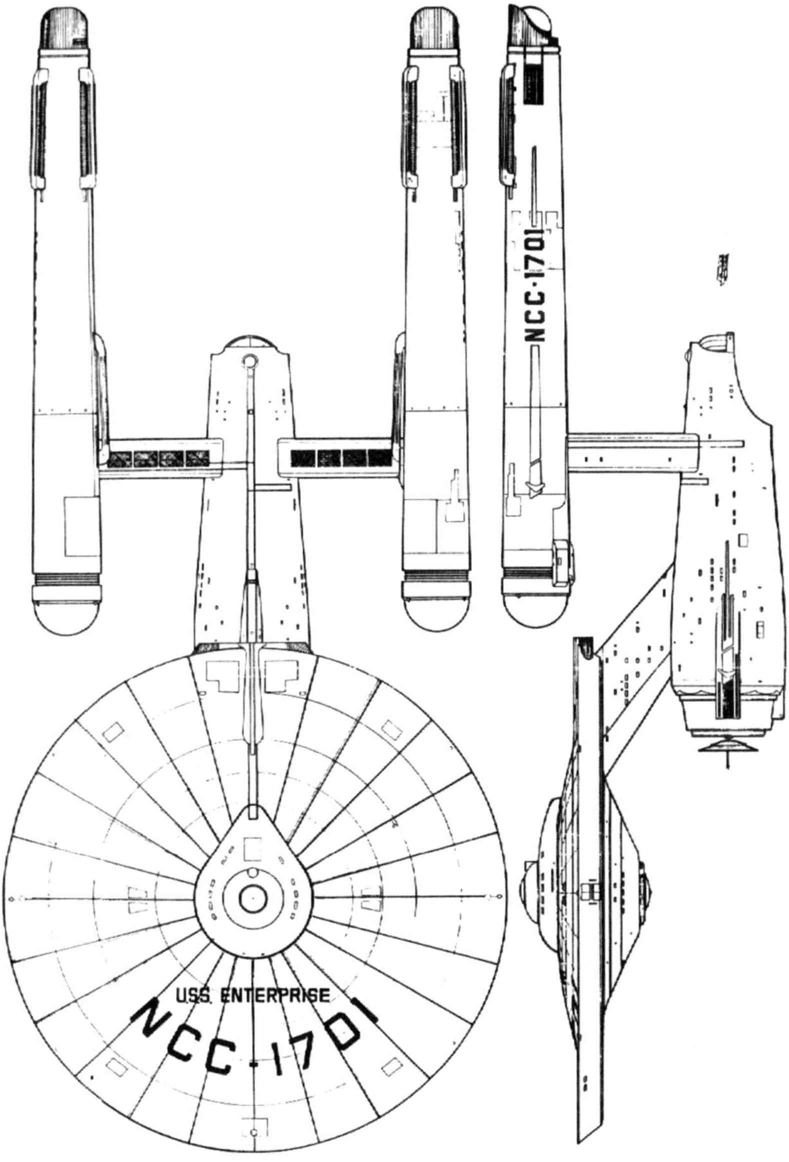

Tracy, and is the primary designer of the NX-01

Diagrams of the 11 foot studio model by Allen Everhart. See AP:04:06.

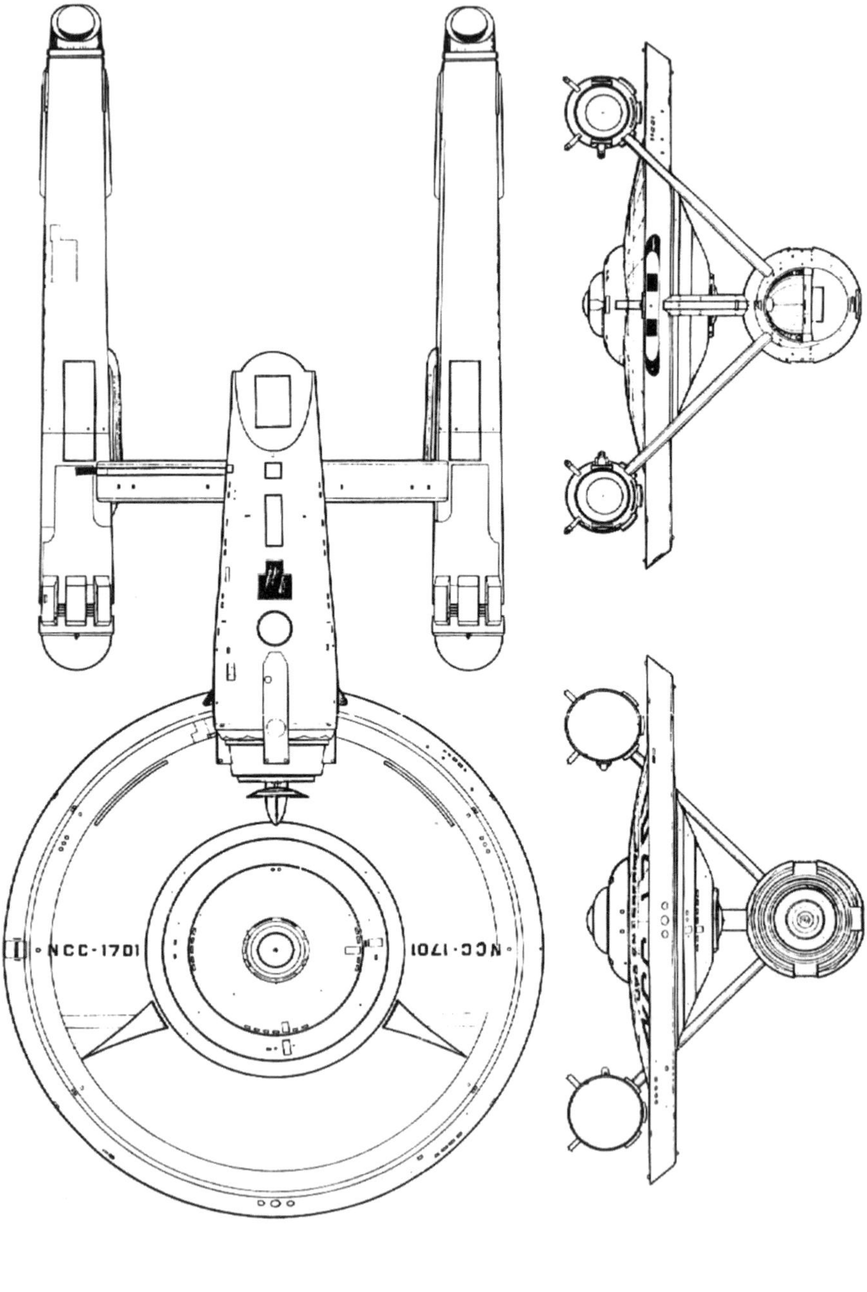
NCC-1701
NCC-1701

StarFleet Assembly Manual 1

ASSEMBLY PROCEDURE FOR HEAVY CRUISER STARSHIPS

a supplemental guide to assembling Heavy Cruiser Starship models

conceived, written and photographed by:
PAUL MATTHEW NEWITT

san francisco yards drawn by:
MARK FINLAY

specially commissioned art by:
AL SIROIS

THIS BOOKLET IS DEDICATED TO THE PEACEFUL USE OF SPACECRAFT PRESENT AND FUTURE

SPECIAL THANKS TO:

Terry Whittier, A.J. Desmaretz, Mark Finlay, Al Sirois, THE SELECTRIC II, Della Newitt, and all the fans who encouraged me to write this booklet.

PRINTED BY A.J. DESMARETZ & CO.

first edition / first printing apr. 1977 CCC
revised / second printing may 1977 M
third printing sept. 1977 M
fourth printing apr. 1978 M

APHC-1 REFERENCE LISTING

OFFICIAL USE ONLY

STAR FLEET DIVISION
SAN FRANCISCO YARDS
UNITED NATIONS / EARTH / SOL SYSTEM
UNITED FEDERATION OF PLANETS

STARDATE 3649.07

PAUL M. NEWITT

1. STARFLEET DIVISION, UNDER DIRECTIVE 7R FROM STARFLEET COMMAND, HAS RELEASED THE ENCLOSED INFORMATION AS YOU REQUESTED OLD CALENDAR DATE NOVEMBER 15, 1976. THIS INFORMATION HAS BEEN DECLASSIFIED BY THE FEDERATION, AND MUST BE "TRANSLATED" INTO ACCEPTABLE COMPREHENSION LEVELS FOR CIVILIANS OF YOUR TIME PERIOD. THIS RELEASE OF INFORMATION HAS BEEN A JOINT EFFORT FROM THE ENGINEERING MICROFILES OF SAN FRANCISCO YARDS AND REFERENCE TEXTS SFAT 44922 AND SFAT MN18 FROM STARFLEET ACADEMY-TERRAN BASE SR-34

2. MOST OF YOUR REQUESTS WERE REGARDED AS CLASSIFIED AND RELEASE TO YOUR TIME PERIOD WOULD ONLY RESULT IN A POSSIBLE CHANGE IN HISTORY. THEREFORE, IT IS NOT POSSIBLE AT THIS TIME TO RESPOND TO YOUR MANY REQUESTS REFERRING TO CLASS I HEAVY CRUISERS. HOWEVER, THE INFORMATION THAT HAS BEEN PROVIDED SHOULD ANSWER MOST QUESTIONS CIVILIANS OF YOUR TIME PERIOD HAVE.

3. IT IS NOT THE POLICY OF STARFLEET DIVISION NOR STARFLEET COMMAND TO RELEASE ANY SUCH INFORMATION. HOWEVER, AN EXCEPTION HAS BEEN MADE IN THIS CASE. OUR PLANETARY AGENTS (FROM YOUR TIME PERIOD) HAVE INFORMED US THAT THE ASSEMBLY OF PLASTIC REPRESENTATIONS OF OUR HEAVY CRUISER STARSHIPS, CLASS 1 MK IV, HAS BEEN TOO DIFFICULT AND GROSSLY INACCURATE FOR CIVILIANS. IT IS THE INTENT OF SFD TO "KEEP AN EYE" ON THE PROGRESS AND COURSE OF OUR PAST - ESPECIALLY THE ASSEMBLY OF OUR STARSHIPS. THEREFORE, WE HAVE PROVIDED INFORMATION TO HOPEFULLY CORRECT THESE PROBLEMS.

4. IT IS THE RESPONSIBILITY OF YOU, A MERE CIVILIAN OF TWENTIETH CENTURY EARTH, TO RE-INTERPRET WHAT WE HAVE PROVIDED. SPECIFIC GUIDELINES OF PRESENTATION HAVE BEEN INCLUDED. YOUR BOOKLET SHALL BE REFERRED TO AS: STARFLEET ASSEMBLY MANUAL ONE - ASSEMBLY PROCEDURE FOR HEAVY CRUISER STARSHIPS. APHC-1/PMN WILL BE THE OFFICIAL CLASSIFICATION OF THE OFFICIAL TWENTIETH CENTURY VERSION. YOU WILL FOLLOW THESE GUIDELINES.

5. IT IS IMPERATIVE THAT YOU KEEP COMMUNICATION TO A MINIMUM. YOUR AIR FORCE (USAF) HAS BEEN INADVERTENTLY RECEIVING OUR SUB-SPACE COMMUNICATIONS. UNTIL WE CAN CORRECT THIS PROBLEM, YOU SHOULD CONSIDER YOURSELF ACTING AS AN INDEPENDENT AGENT. AS THEY SAY IN YOUR TIME PERIOD - "GOOD LUCK".

PEACE TO OUR PAST

ℓαξω ∂˜ ~εδιττ

CHIEF OF OPERATIONS, SFD
SAN FRANCISCO YARDS

{END OF COMMUNIQUE}

StarFleet Division

(Response to Communiqué S.D. 3649.07 shown on previous page)

...THE TWENTIETH CENTURY TERRAN VERSION OF STAR FLEET DIVISION IS PROUD AND HONORED TO PROVIDE THE RELEASED INFORMATION PERTAINING TO THE ASSEMBLY OF HEAVY CRUISER STARSHIPS. THE FIRST PHASE OF TRANSLATION HAS BEEN COMPLETED AND IS NOW AVAILABLE TO TWENTIETH CENTURY CIVILIANS. VARIOUS PARTS OF THE RELEASED INFORMATION BECAME TOO ADVANCED TO "TRANSLATE" AND WILL REQUIRE FURTHER IN-DEPTH STUDY. FUTURE RELEASES OF THE ADVANCED TECHNIQUES MAY BE AVAILABLE AT A FUTURE DATE. STARFLEET ASSEMBLY MANUAL 1 SHOULD PROVIDE ADEQUATE, BASIC INFORMATION AT THIS TIME.

Fans, Modelers, Cadets:

Your StarFleet Assembly Manual One is just the first of a series of supplemental booklets about the assembly and modification of Federation and Alien spacecraft models. Binder holes have been provided so that several booklets can be collected and kept neatly in a three-ring binder (8½"x5½"). This first booklet is aimed to provide basic, helpful information to those who may not be into model building, but wish to obtain fine results with little effort. Most fans find after building their first starship, that something doesn't look right about it. It is the intent of this booklet to clear up most of these problems. There are also some hints and techniques for those who are very much into model building. Some of you may only need this information to get your starship together, while some of you may find this only a beginning to what can be done to create realistic starship models. There are unlimited possibilities. StarFleet Division wishes to eliminate the "hassle" in starship model assembly, and provide accurate information to obtain satisfying results.

This single effort isn't expected to solve every problem, but it is a start. If you have any specific questions, problems, suggestions, or comments, please write. Communication is the beginning of understanding.

Peace and Long Life,

Paul M. Newitt
Author, S.A.M. 1

/pmn

P.S. Please enclose SSAE when writing.

SAN FRANCISCO YARDS - GEN. DESCR.

PICTURED BELOW (FIG. 1) IS A GENERAL TOPOGRAPHIC PROJECTION OF QUADRANT CAL-94, 95, 93, WHICH INCLUDES SAN FRANCISCO YARDS. ROUGHLY TWO HUNDRED YEARS AGO IT WAS KNOWN AS "NORTHERN CALIFORNIA" OR "THE BAY AREA". FEW ARTIFACTS REMAIN OF THAT ERA DUE TO THE WIDESPREAD DESTRUCTION BY THE EUGENICS WARS OF THE LATE 1990's.

THIS COMPLEX IS DESIGNED SPECIFICALLY FOR THE ASSEMBLY OF PRE-FABRICATED SECTIONS OF HEAVY CRUISER STARSHIPS. THE YARDS ARE LOCATED 18.8 KILOMETERS S-SE FROM THE ACADEMY AND ARE DIRECTLY EAST OF THE SAN FRANCISCO METRO SECTOR.

DURING THE LATE TWENTIETH CENTURY, AND BEFORE THE QUAKE OF 2004, THIS SAME AREA WAS OCCUPIED BY THE HUNTER'S POINT NAVAL SHIPYARD (U.S. NAVY), AS SHOWN IN FIG. 2.

(figure 1)

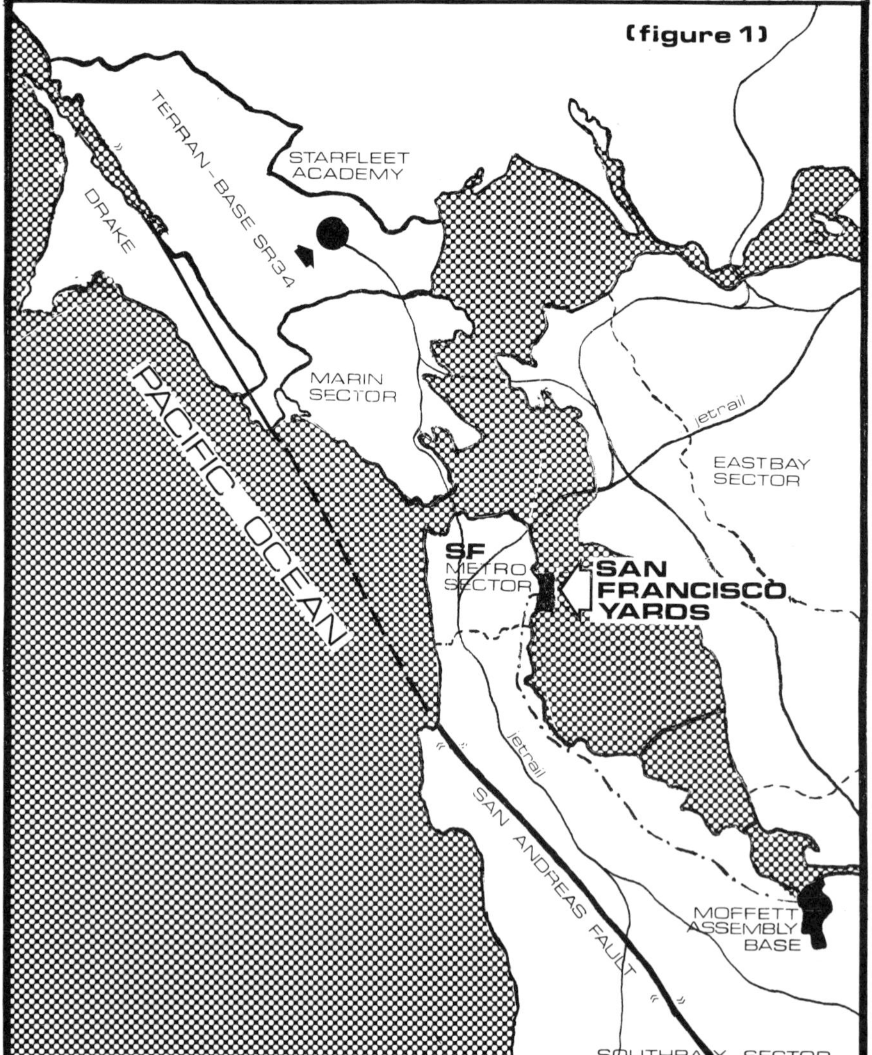

SAN FRANCISCO YARDS-GEN. DESCR.

AS CAN BE SEEN IN FIG. 3, THE AREA OF LAND SURROUNDING THE YARDS (THEN SUBMERGED) ROSE DURING THE QUAKE OF 2004 PROVIDING NEW AND MORE STABLE LAND ON WHICH THE YARDS ARE SITUATED TODAY. SINCE THE DEVELOPMENT OF TERRATONICS, THE FEAR OF SEVERE EARTHQUAKE DAMAGE HAS BEEN GREATLY REDUCED.

SOUTH OF THE YARDS ARE FACILITIES DESIGNED TO ACCEPT SECONDARY COMPONENTS TO BE OUTFITTED AFTER BASIC HULL ASSEMBLY IN SPACE, JETRAILED IN FROM MOFFETT ASSEMBLY BASE. THESE ITEMS INCLUDE: LIFE-SUPPORT SYSTEMS, COMMAND AND ENGINEERING CONSOLES, IMPULSE POWER COMPONENTS, MEDICAL EQUIPMENT AND FURNISHINGS, SHUTTLECRAFT MAINTENANCE EQUIPMENT, ETC.

STARFLEET DIVISION'S MOFFETT ASSEMBLY PLANT IS LOCATED 38.4 KILOMETERS SOUTHEAST OF SAN FRANCISCO YARDS, AND WAS FORMERLY OCCUPIED BY NASA-AMES RESEARCH CENTER.

LOCATED WITHIN THE MAIN YARDS IS WHERE PRIMARY AND SECONDARY HULL SECTIONS ARE CONSTRUCTED (FIG. 3). MAIN PROPULSION (WARP ENGINES) IS CONSTRUCTED IN THE COMPLEX AS SHOWN. EACH MAJOR PART OF A STARSHIP IS ASSEMBLED IN "SHELL" FORM AND BEAMED INDEPENDENTLY INTO A PRE-DETERMINED POINT IN SPACE. EACH SEPARATE SECTION CONTAINS A TRANSPORTER (TRAN 2, TRAN 3) CAPABLE OF TRANSPORTING 5200 METRIC TONS RESPECTIVELY (SIMILAR TO TRANSPORTER AS DESCRIBED IN TM:379260-3/03:16:50).

THE MAIN YARDS OCCUPY AN AREA APPROXIMATELY 3.42 SQUARE KILOMETERS. THE COMPLEX EXTENDS DOWNWARD TO FOUR UNDERGROUND LEVELS. POWER TO THE PLANT IS SUPPLIED JOINTLY BY SOLAR ABSORPTION AND KINETIC HYDROLOGY AND IS CAPABLE OF SELF-SUPPORT AS OUTLINED BY FLEET OPERATIONS REQUIREMENTS - STAR FLEET COMMAND. FOR FURTHER DETAILS REFER TO SFAT 44922.

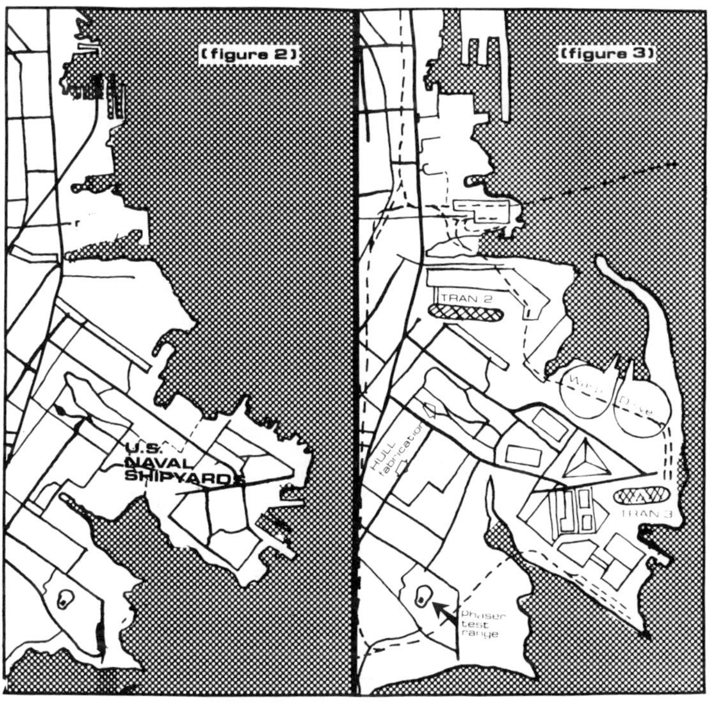

PRIMER 1: tools & materials.

TO ASSEMBLE YOUR AMT ENTERPRISE YOU WILL REQUIRE A MINIMUM OF TOOLS AND MATERIALS WITHOUT A MAJOR INVESTMENT. THE BASIC ASSEMBLY OF YOUR STARSHIP WILL ONLY REQUIRE TOOLS THAT ARE INEXPENSIVE AND READILY AVAILABLE. Given below are several lists of tools and materials, some of which are optional. Some additional tools and materials from the optional list may make assembly easier. The use of all of these items will be referred to in the text. Most all of them can be found in any hobby, toy, or artist's shop.

List #1 (minimum:beginner):
- 1 - emery board
- 1 - single edged razor blade
- 1 - packet of sandpaper, fine to extra fine (Flexi-grit)
- 1 - paint brush, fine tip
- 1 - tube Testor's Plastic Cement
- 1 - small roll of masking tape (3/4" width)
- 1 - pair tweezers

List #2 (optional tools):
- 1 - X-acto (or X-po) knife, small handle
- 1 - assortment of X-acto blades
- 1 - bottle Testor's or Pactra Brush-on Plastic Cement
- 1 - tube Testor's Contour Putty for plastic models
- 1 - tube "Crazy Glue" (CYANOACRYLATE ADHESIVE)
  <u>CAUTION</u>: CRAZY GLUE SHOULD ONLY BE USED WITH ADULT SUPERVISION.
- 1 - broad tip paint brush, artist's brush preferable
- 1 - bottle Champ Decal Set
- 1 - bottle (Micro-scale) - Micro-mask
- 1 - roll graphic tape (1/8" wide) in chrome, white or black (or model railroad decals of stripes)
- 1 - can matte or semi-matte lacquer spray, such as Testor's Dullcote or photo-matte spray
- 1 - sheet dry transfer lettering; microgramma extended or eurostile extended, 20 pt. caps, letters are 7/32" high

List #3 (optional materials):
These items are optional additions you may want to add, depending upon the method of display you may choose.
- 1 - Plastruct rod (3/16" diameter, 8-10" long)
- 1 - brass tube, slightly larger diameter than 3/16", 8-10" long
- 1 - round, square, or oval wood base, similar to those used for découpage. Must be a minimum of 24 sq. in.
- 1 - can wood stain for base, light or dark

If any items have been omitted, they may appear in the text. Please scan the ASSEMBLY PROCEDURE.

PRIMER 2: painting & colors.

IF YOU PLAN TO PAINT YOUR HEAVY CRUISER STARSHIP, BELOW ARE SOME HINTS TO HELP YOU DETERMINE WHAT COLORS TO USE.

1. Heavy Cruiser Starships are not white, and never were. The appearance of white is from the bright lights reflecting off of the actual color -- grey-platinum. These lights that the studios used "washed-out" the grey to appear white.

2. The production studios painted their models (one 14 ft. and one 4 ft.) with a white base, covered with a silvery coating. During the second year of production, 1967, the 14 ft. model was completely repainted, weathered, and included the addition of those white "balls" at the rear of the engines.[1] The new color of grey-platinum (or neutral grey) was used to accept special mood lights that created an appearance of blue to purple. (Please refer to Terran Based References at the end of this booklet for information in obtaining color photos of the studio models). Movie film and your TV's color tint distorted the actual color even further.

3. The reflected light in outer space greatly differs from that on earth because of the lack of an atmosphere. The Earth's atmosphere allows all colors of white light to filter into the color spectrum as we know of it. Often the colors of NASA spacecraft, such as white and black, appear differently in space. White becomes grey and black becomes charcoal. For example, on a night with a full moon, take a color wheel outside and compare the colors as they appear outside to that of a lighted room. You will find that the actual color is distorted by the absence of the white light spectrum. Grey-platinum is a color that absorbs and reflects all colors, often hiding the true color. That is why the ENTERPRISE appears to be many different colors.

4. If you choose to paint your starship a color that you have seen on television, try to add a very small amount of the desired color to a lightened grey-platinum. For example, if you like a blue tint, add 3 parts blue to 10 parts light grey, that has had 1 part of silver or platinum already added. If you wish to do this with aerosol spray paint, start with a light grey (sea grey) base first, then try a color, such as blue or purple, in a translucent paint form.* Spray this at a further than usual distance of 20-25 inches. This will make the paint "mist" on top of the grey and give it a dramatic, in-space look. If you are unsure of the color or technique, you should try experimenting on another piece of white styrene plastic first.

5. Below are listed most all of the colors you may require. Four scale paint manufacturers are listed to provide easy availability. For beginners, the Pactra Military Paints produce fine results with little effort. Testor's Paints are

1. From TREK #5. See References Page.

* - Explained in following text on next page.

listed as an alternative to Pactra, and include the translucent spray paints for special effects.* Floquil Paints are intended for Model Railroads, but can be used just as effectively on starships. Humbrol Paints are especially mixed to match military models. IF you are an advanced modeler, consider Humbrol as a perfect match to those colors the studios used on the 14 ft. model. Both Floquil and Humbrol Paints, especially when airbrushed, produce professional results. MORE DETAILS ON PAINTING TECHNIQUES will be in the ASSEMBLY PROCEDURE.

PACTRA MILITARY FLATS

(For Beginners)

- M 1 Scale Black
- M 2 Primer White
- M 4 Aero Blue
- M 8 Brick Red
- M11 Fabric Yellow
- M13 Camouflage Gray
- M15 Steel

Spray

- SM 13 Camouflage Gray - Lt. Gray
- SM 4 Aero Blue
- SM 12 Rebel Gray - Dk. Gray

FLOQUIL (Model Railroad Scale Paint)

- RR 11 Reefer White
- RR 9 Primer
- RR 10 Engine Black
- RR119 Graphite
- RR 85 Antique White
- RR 56 Great Northern Sky Blue
- RR100 Old Silver
- RR144 Platinum Mist
- RR 17 Weathered Black
- RR253 Weathering Paste
- RR 5 Glaze
- RR 3 High Gloss
- RR 1 Dio Sol (Thinner)

Plastic may be treated with "barrier" before applying Floquil Paint.

NOTE: Not all of the colors listed above will be required. Some colors are optional. PLEASE REFER TO PAINTING SCHEMATICS AT THE END OF THE ASSEMBLY PROCEDURE.

TESTOR'S

- 50 Flat Red
- 49 Flat Black
- 38 Gray
- 37 Blue Gray
- 14 Yellow
- 62 Flat Sky Blue
- 46 Silver
- 38 Gray (Spray)
- 58 Flat White (Spray)
- 12D Dullcote (Matte Spray)
- * Translucent colors

HUMBROL MILITARY COLORS

- HU.10 Grey 36622†
- H 1.5 Grey†
- HJ.2 Grey A/NZ††
- HN.1 Lt. Grey††
- HB.6 Sea Grey
- HU.6 Lt. Grey
- HB.3 Ocean Grey
- HG.3 Hellgrau 76
- HN.2 Dk. Grey
- HG.4 Dk. Grey
- HJ.6 Silver
- HB.14 Aircraft Silver
- HG.1 Schwaezgrun 70
- HU.12 Night Black
- HU.11 Airframe White
- H1.6 Insignia White

† - produces near perfect match to studio models (cold grey)

†† - same as above (warm grey)

ASSEMBLY PROCEDURE: HC-1

Before you begin to assemble your Heavy Cruiser Starship, remember that it should not be completed in the same day that you begin. Ninety-nine percent of the disappointments and discouragements from a completed starship are from doing things too fast. A series of evenings or two to three weekends would produce better results to anyone, no matter how inexperienced they may be. Common problems like "sagging warp engines", and primary hulls that detach themselves from the rest of the ship should be eliminated by following this assembly procedure carefully.

You will find that your patience and time will pay off. No longer will you have to stuff your starship model under the bed or be ashamed of it when your friends come over. You will be proud of your starship model, even if it is your first attempt at any model, <u>only</u> if you <u>carefully</u> read the instructions.

AMT Corporation, the company who manufactures this model, has done a fine job of revising the kit and its assembly. Since 1968 it still has retained its popularity (even to Klingon ships, a close second). Three versions of attaching the engine pods to the secondary hull have been attempted by AMT over the years, a sign that they respond to modelers and their problems of "sagging warp engines". The latest version released approximately February, 1976, is their most accurate revision. A new fool-proof way of attaching the warp engines has been added as well as slightly shorter and tapered nacelles, with "balls" on the rear. The main sensor assembly attaches easier, as well as the dome assembly at the front of the nacelles. AMT Corporation has responded very well to the development of accurate and easy to assemble Heavy Cruiser Starships.

It is important that you go over these supplemental instructions along with those that come with your kit, to decide upon what you want the end result to look like. You also must decide how much effort you want to put into it, i.e. if you are going to paint it, and if so, what colors, if you are going to add any modifications, and how you intend to display it.

Here are some other points to remember:

1. This booklet is only a supplement to the AMT instructions. The order of assembly may be different, but the diagrams that AMT provides may illustrate certain parts of assembly in a better way.

2. Be careful when using plastic cement. It has an acetone base and therefore, eats into plastic. Excess plastic cement should be avoided because it will only ruin your efforts. A minimum amount of cement and contact pressure should be adequate. On broad areas of contact, like the primary hull for example, use a cotton swab with the loose cotton removed to spread the plastic cement evenly.

3. If you are using Crazy Glue, be <u>extremely careful</u>, and if you are under 12 years, please have <u>adult supervision</u>. It can be very dangerous!

4. If you sand all seams and points of contact on the ship, and take your time in alignment of certain parts, your ship will look surprisingly realistic.

5. PLEASE READ THESE INSTRUCTIONS AND THE ENTIRE ASSEMBLY PROCEDURE BEFORE BEGINNING ASSEMBLY!

BEGINNING

1. AFTER OPENING THE BOX, YOU WILL DISCOVER A PLASTIC BAG FILLED WITH PLASTIC PARTS. ALSO IN THE BOX ARE THE DECALS, INSTRUCTIONS, AND THE CLEAR-GREEN DOMES AS INDICATED BY THE ARROW. THERE SHOULD BE 4 GROUPS OF PARTS HELD ONTO "SPRUES"; BOTH HALVES OF THE PRIMARY HULL (SAUCER SECTION) AND THE OTHER 2 HALVES OF THE ENGINE PODS (NACELLES). PLEASE CHECK THE INSTRUCTIONS THAT COME WITH THE KIT, AS WELL AS THE PHOTO AT THE RIGHT TO MAKE SURE THAT EVERYTHING IS THERE. IF SOMETHING IS MISSING OR NOT MOLDED RIGHT, TAKE THE KIT BACK TO THE STORE WHERE YOU PURCHASED IT, OR WRITE AMT EXPLAINING THE PROBLEM. IN EITHER CASE, YOU SHOULD BE ABLE TO GET THE MISSING OR BROKEN PART(S).

2. THE NEXT THING TO DO IS TO REMOVE THE PARTS FROM THE SPRUES (PLASTIC HOLDERS ATTACHED TO PARTS). LARGER PARTS CAN BE REMOVED BY FLEXING BACK AND FORTH AS INDICATED. EACH PART SHOULD RELEASE EASILY.

3. SMALLER PARTS CAN BE REMOVED FROM THE SPRUES BY USING A SHARP KNIFE. THESE SAME PARTS CAN ALSO BE REMOVED BY FLEXING, USING A KNIFE WILL GET RID OF ROUGH SPOTS ON THE ENDS AND EDGES. ON THE INTERCOOLERS, <u>DO NOT</u> CUT OFF THE GUIDE PINS AS SHOWN BY THE ARROW. (THESE GUIDE PINS GO INTO THE HOLES ON THE NACELLES).

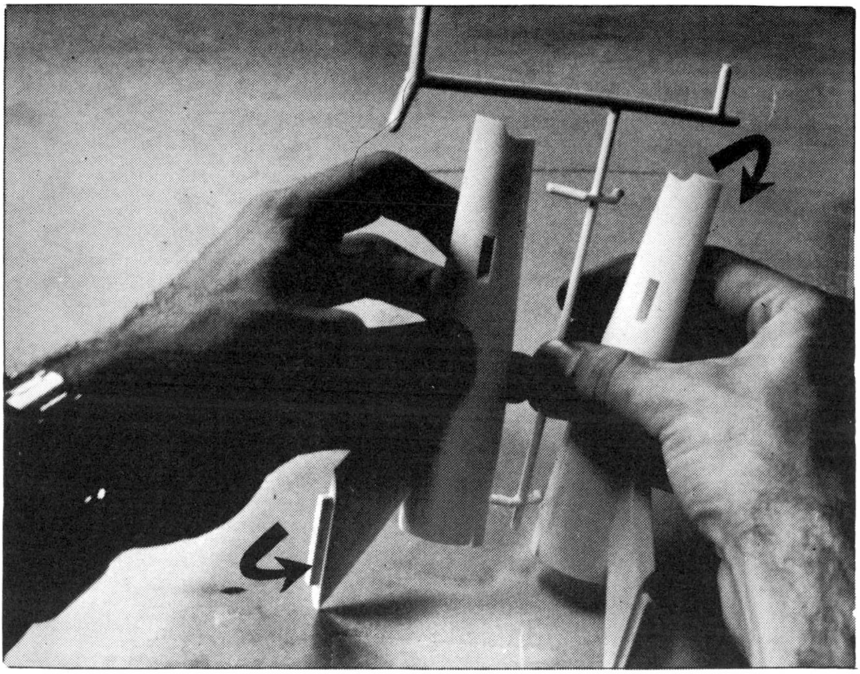

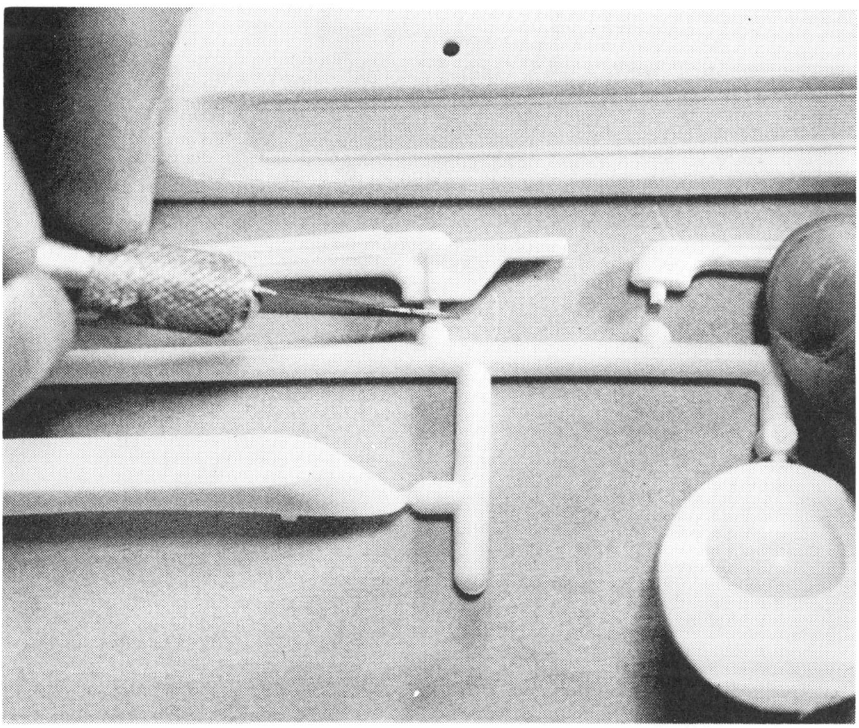

4. AFTER YOU HAVE REMOVED ALL OF THE PARTS FROM THE SPRUES, SAND ALL OF THE ROUGH PLACES WITH VERY FINE SANDPAPER, FILE, OR EMERY BOARD. YOU MAY NEED TO USE A SHARP POINTED KNIFE ON ROUGH PLACES AS SHOWN IN THE BOTTOM PHOTO. AFTER SANDING ALL OF THE PARTS TO REMOVE EXCESS PLASTIC, THEY SHOULD FIT BETTER. SHOWN ARE THE 3 GROUPS THAT WILL BE ASSEMBLED:

GROUP 1. THE PRIMARY HULL (SAUCER SECTION)

GROUP 2. THE SECONDARY HULL

GROUP 3. NACELLES (WARP ENGINES)

PLEASE TEST-FIT ALL THE PARTS TO FAMILIARIZE YOURSELF WITH HOW THEY GO TOGETHER. SOME PARTS MAY REQUIRE SANDING, SOME PARTS MAY NOT.

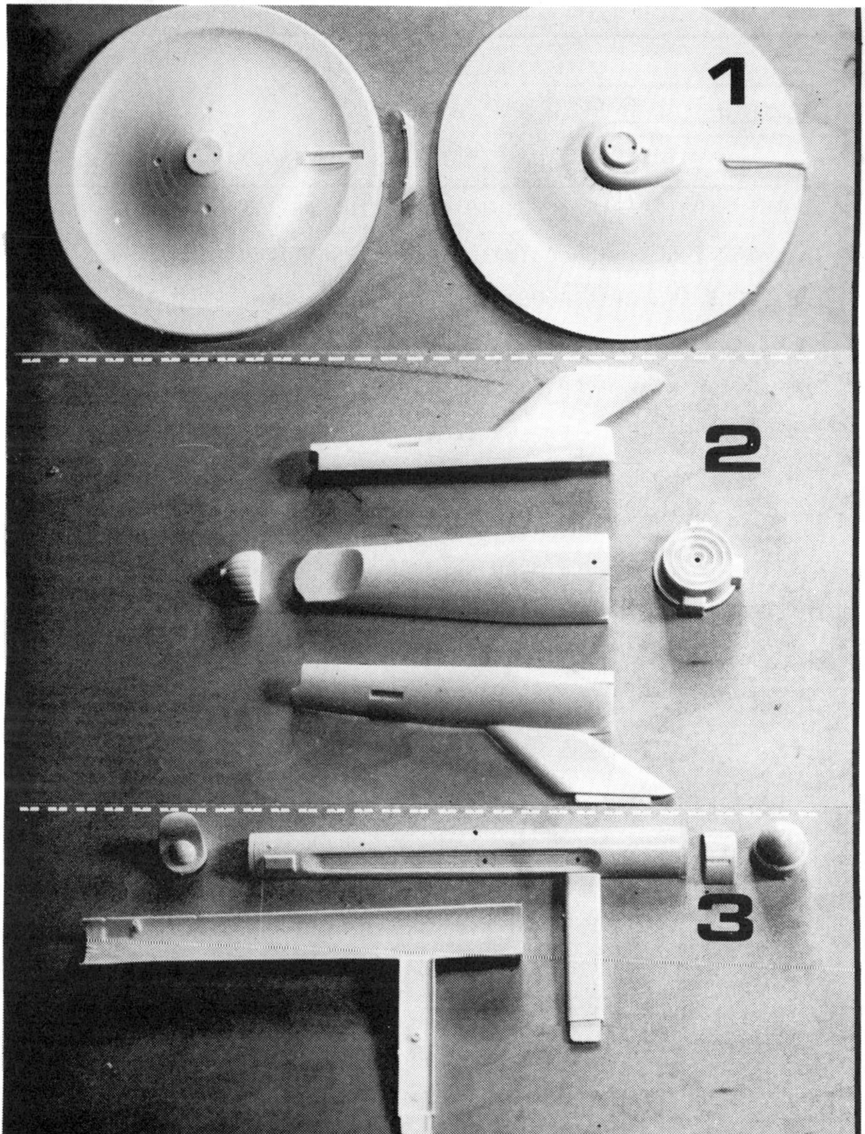

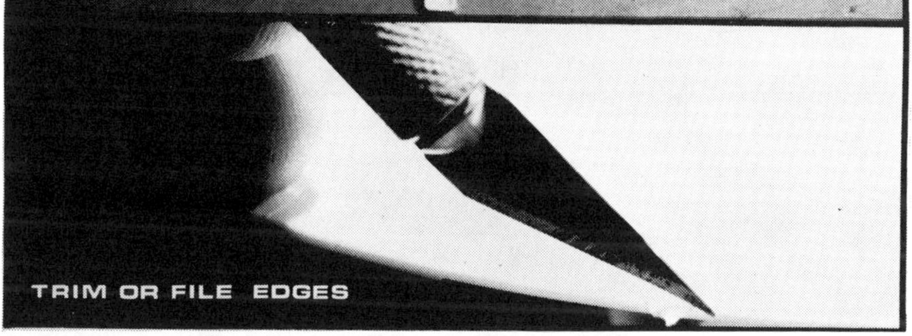

TRIM OR FILE EDGES

NACELLES

5. THE FIRST PARTS TO BEGIN CEMENTING ARE THE NACELLES. VERY CAREFULLY, RUN A THIN LINE OF PLASTIC CEMENT DOWN THE EDGES (WHERE ARROWS SHOW) AND APPLY CEMENT ON THE HOLE INSIDE THE STRUT, AS WELL AS ON THE GUIDE PINS AT THE INSIDE REAR. APPLY CEMENT ONLY TO ONE HALF OF THE 2 NACELLE PIECES. EXCESS CEMENT DOES NOT INCREASE HOLDING POWER AND WILL ONLY DELAY DRYING TIME. WHEN BRINGING BOTH HALVES TOGETHER, FIRST APPLY PRESSURE TO THE STRUT SECTION. YOU SHOULD HEAR A "SNAP" SOUND INSIDE. THIS IS THE GUIDE PIN AND HOLE OF THE PYLON MAKING CONTACT. NOW THE 2 HALVES SHOULD BE TOGETHER. NEXT IS THE ALIGNMENT OF THE NACELLE ENDS SO THAT THEY FORM AN EVEN CIRCLE. THE BOTTOM PHOTO SHOWS A NACELLE END BEFORE ALIGNMENT. BY FORCING THE UNALIGNED END IN AN ALIGNED POSITION AS THE GLUE IS DRYING, IT SHOULD CONFORM TO FORM AN EVEN CIRCLE. MASKING TAPE SHOULD ALIGN THE EDGES EVEN FURTHER, MAKING SURE IT IS APPLIED TIGHTLY. PLACING A CYLINDRICAL OBJECT INTO EACH END, SUCH AS A CORK, SHOULD ALSO HELP ALIGNMENT, BUT IS NOT REQUIRED. AFTER YOU HAVE TAPED THE NACELLES AND THEY ARE REASONABLY ALIGNED ON THE ENDS AND EDGE, PLACE THEM ASIDE AND ALLOW THEM TO DRY FOR AT LEAST 4 - 6 HOURS. AFTER THEY HAVE DRIED, REMOVE THE TAPE AND SAND OR FILE THE SEAMS UNTIL THEY DISAPPEAR.

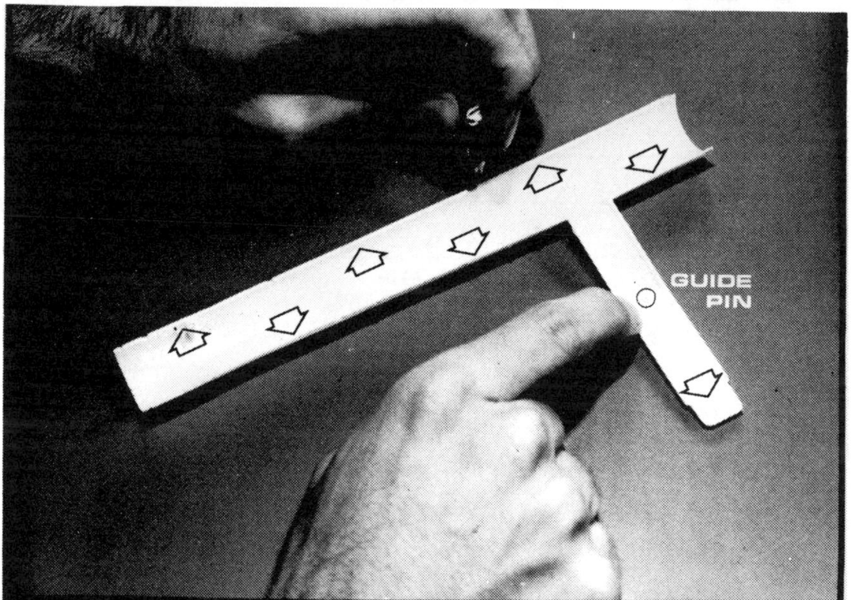
GUIDE
PIN

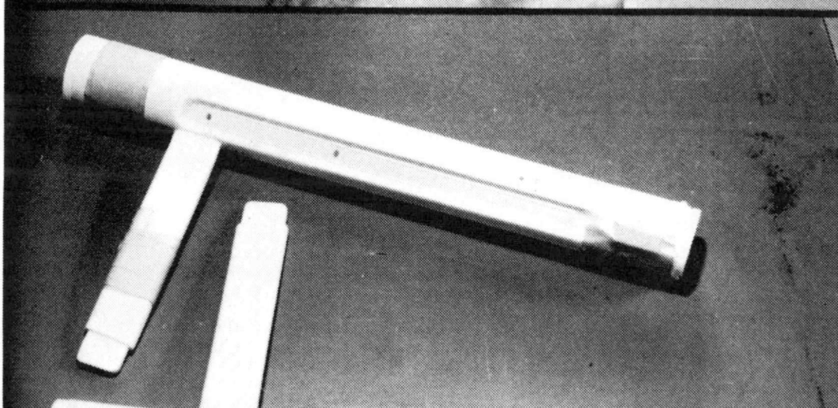

ALIGN

SECONDARY HULL

6. AFTER THE NACELLES HAVE DRIED, PROCEED TO THE SECONDARY HULL. START BY MAKING SURE THAT THE EDGES TO BE CEMENTED ARE FREE OF EXCESS AND ROUGH PLASTIC. THIS CAN BE DONE BY USING VERY FINE SANDPAPER, AN EMERY BOARD, OR SMALL FILE. ONCE THE EDGES SEEM SMOOTH, WITHOUT CEMENT PLACE TOGETHER THE 2 HALVES AS SHOWN ON THE FOLLOWING PHOTO SUPPLEMENT PAGE. THE 2 HALVES SHOULD GO TOGETHER WITHOUT ANY GAPS, CRACKS, OR OBVIOUS SEAMS. AFTER YOU FEEL THAT YOU ARE READY FOR CEMENTING, START BY ONLY APPLYING CEMENT TO THE DORSAL SECTION (SEE PHOTO). APPLY CEMENT TO ONLY 1 HALF, AND BRING THE OTHER HALF INTO PLACE. THE CEMENT WILL NOT DRY INSTANTLY, SO YOU HAVE A FEW SECONDS TO POSITION THE TWO PIECES INTO PLACE. EXCESS CEMENT WILL ONLY DELAY DRYING TIME. YOU WILL NOTICE THAT THE OTHER END OF THE 2 HALVES OF THE SECONDARY HULL MAY NOT BE TOGETHER AND ALIGNED RIGHT. THIS WILL BE CORRECTED LATER. FIRST, ALLOW THE DORSAL SECTION TO DRY BY ADDING SOME MASKING TAPE, AS SHOWN IN THE LOWER PHOTO.

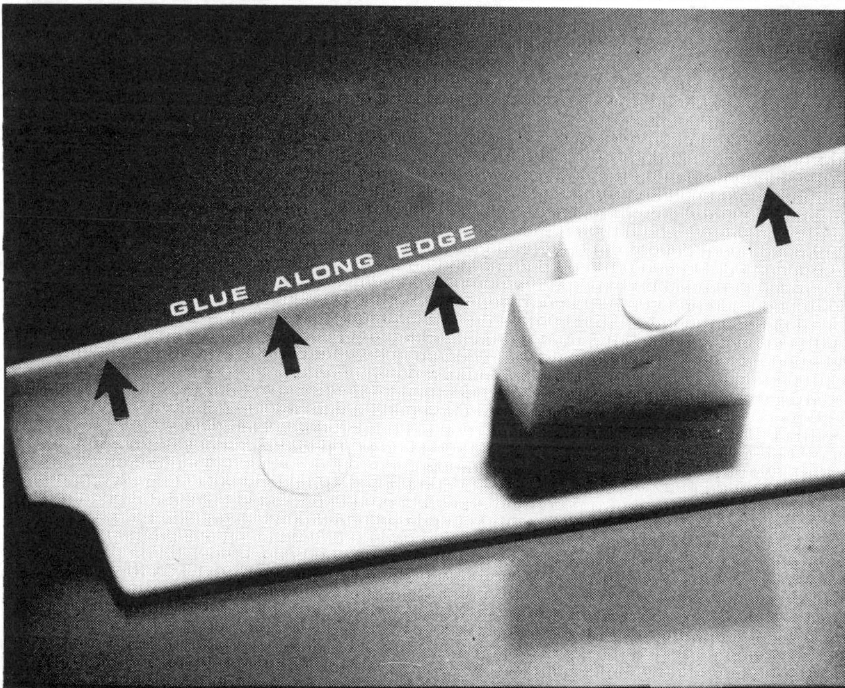
GLUE ALONG EDGE

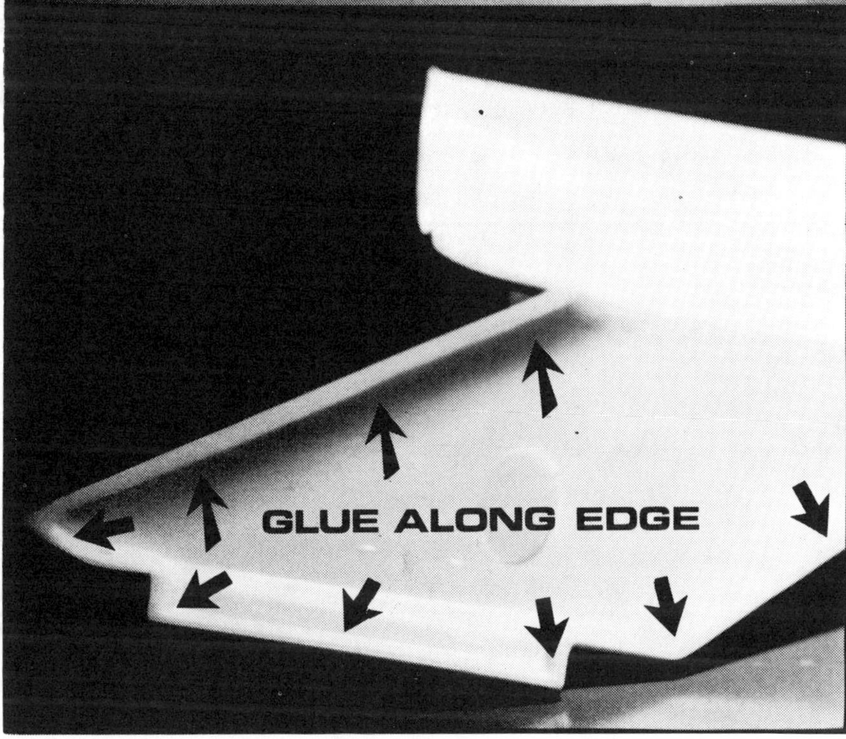
GLUE ALONG EDGE

7. AFTER A FEW HOURS, YOU MAY CONTINUE JOINING THE SECONDARY HULL BY PLACING YOUR FINGERS (OR A CLAMP OR PLIERS) ON THE REAR, ABOVE THE SHUTTLECRAFT BAY HANGAR DOORS. ON THE INSIDE APPLY A LIGHT AMOUNT OF CEMENT TO THE SEAM. PLACE A THIN PIECE OF CARDBOARD OR PAPER (ABOUT 1/8" THICK) BETWEEN THE NACELLE STRUT INSERTS SO THAT THEY DO NOT TOUCH EACH OTHER. APPLY MASKING TAPE TO THE OUTSIDE OF THE SEAM AS SHOWN BY THE ARROWS. ALLOW TO DRY FOR 4 - 6 HOURS. THIS PART MAY BE DIFFICULT, SO TAKE YOUR TIME. MINOR MISTAKES IN CEMENTING CAN BE SANDED OFF LATER.

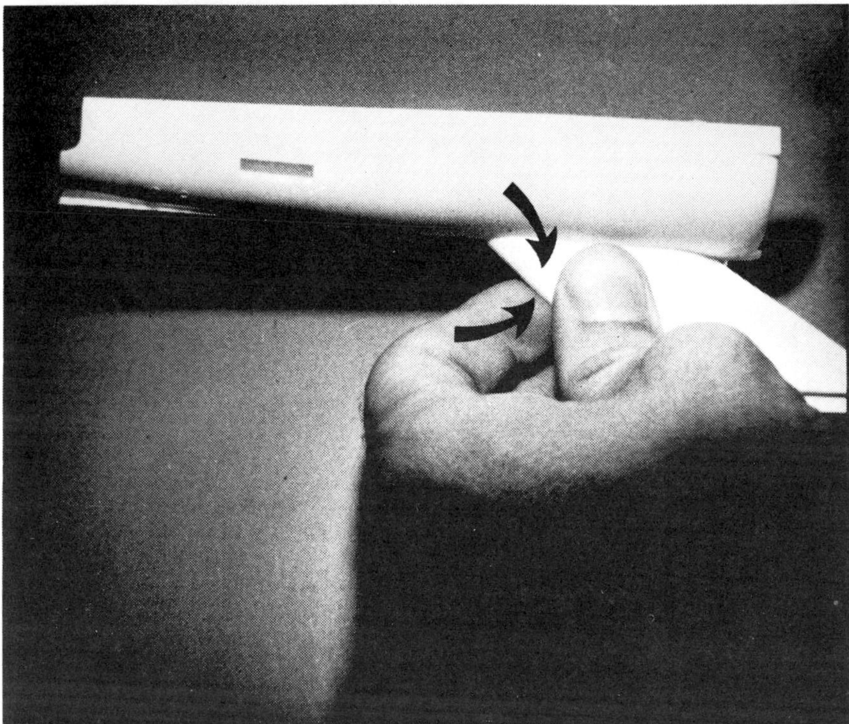

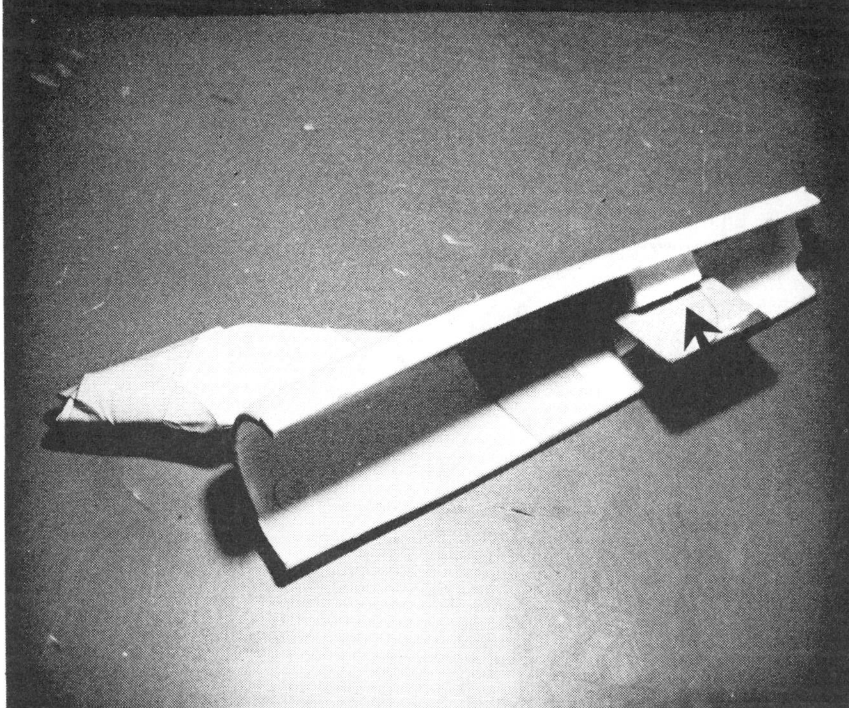

8. WHILE THE UPPER SECTION OF THE SECONDARY HULL IS ASIDE DRYING, THE HANGAR DOORS CAN BE CEMENTED TO THE REAR BOTTOM SECTION OF THE SECONDARY HULL.

9. WHEN THE UPPER AND LOWER SECTIONS OF THE SECONDARY HULL HAVE DRIED, TRY TO TEST-FIT THE 2 SECTIONS TO CHECK FOR ANY GAPS AND LARGE SEAMS. IT MAY BE NECESSARY TO TRIM AND/OR SAND THE TOP EDGE OF THE HANGAR DOORS TO PROVIDE A BETTER FIT. CONTINUE TRIMMING AND SANDING UNTIL ALL SEAMS ARE BARELY VISIBLE.

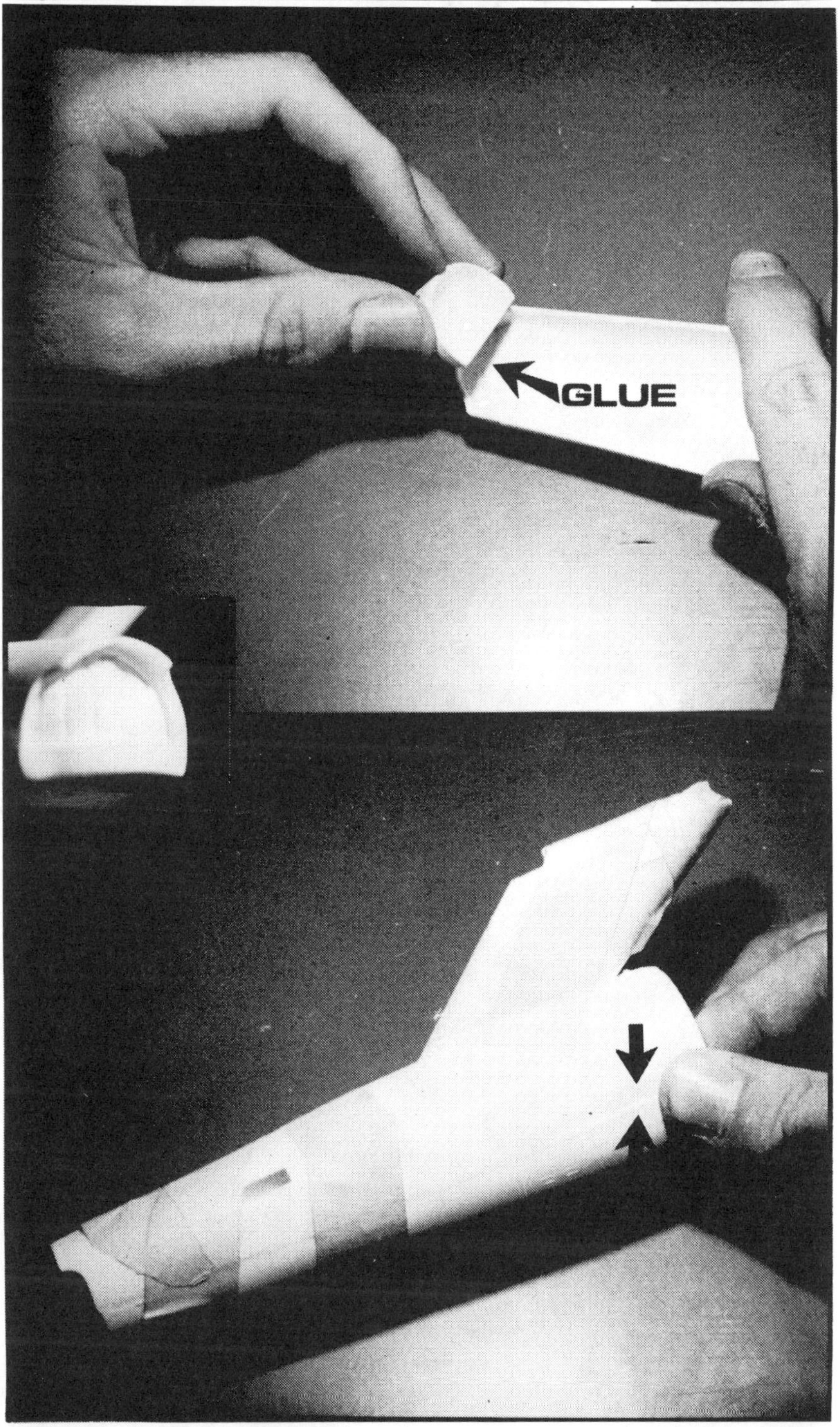
GLUE

SFO 4
SFO 4

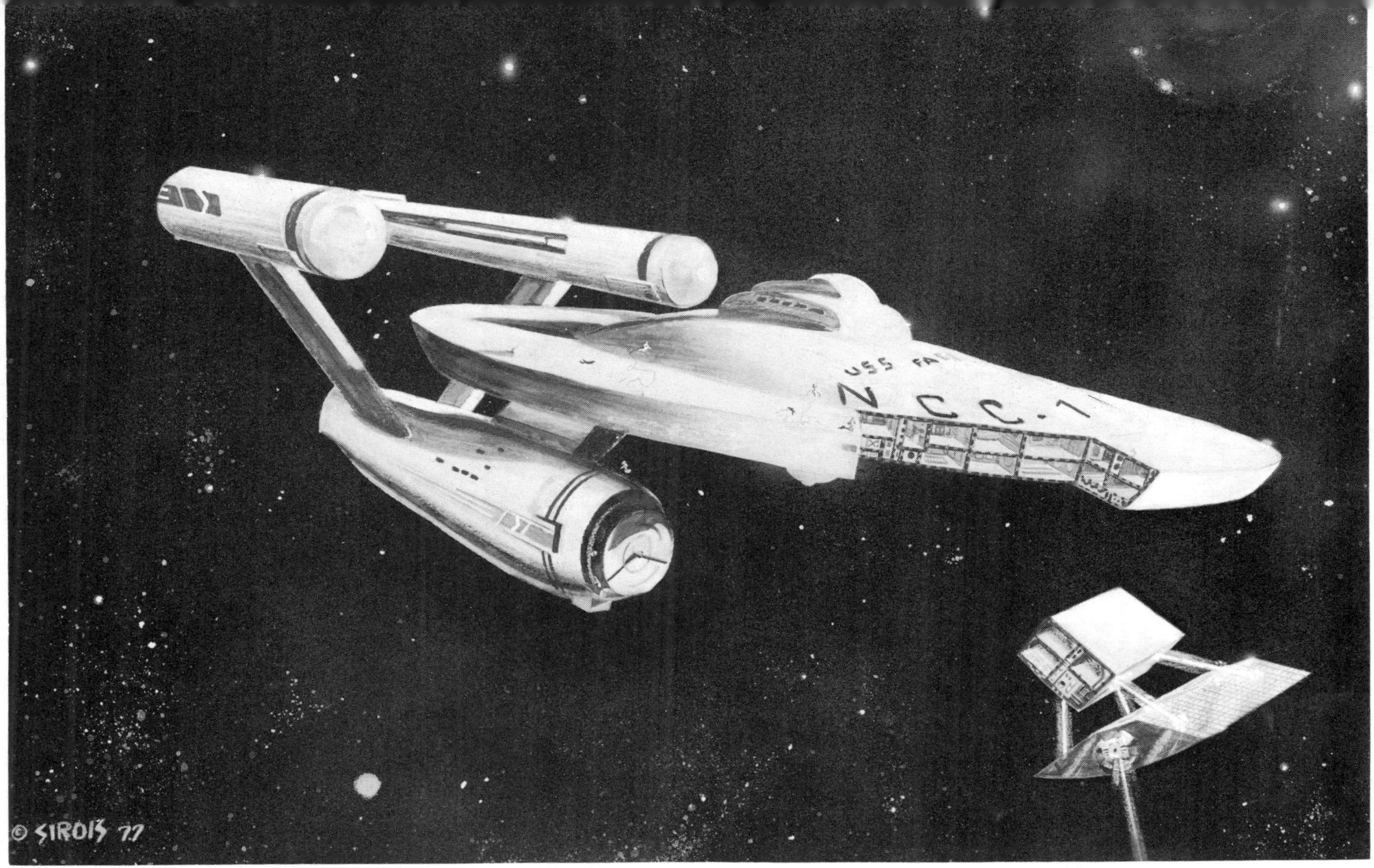
© SIROIS 77

10. WHEN YOU ARE READY TO JOIN BOTH SECONDARY HULL SECTIONS TOGETHER, APPLY CEMENT TO THE LOWER SECTION (WHERE ARROWS SHOW) AND POSITION INTO PLACE. THERE ARE SMALL GUIDES ON THE INSIDE EDGE OF THE BOTTOM SECTION TO HELP THE POSITIONING. (DON'T SAND THEM OFF). IF YOU ARE HAVING A DIFFICULT TIME ALIGNING THE EDGES, AND THE SEAMS ARE STILL VISIBLE AFTER CEMENTING, DON'T FEAR! THERE ARE WAYS TO CORRECT THIS, AS WILL BE EXPLAINED LATER. TAPE THE JOINED SECTIONS TO ENSURE PROPER ALIGNMENT AND ALLOW THIS TO DRY 4 - 6 HOURS.

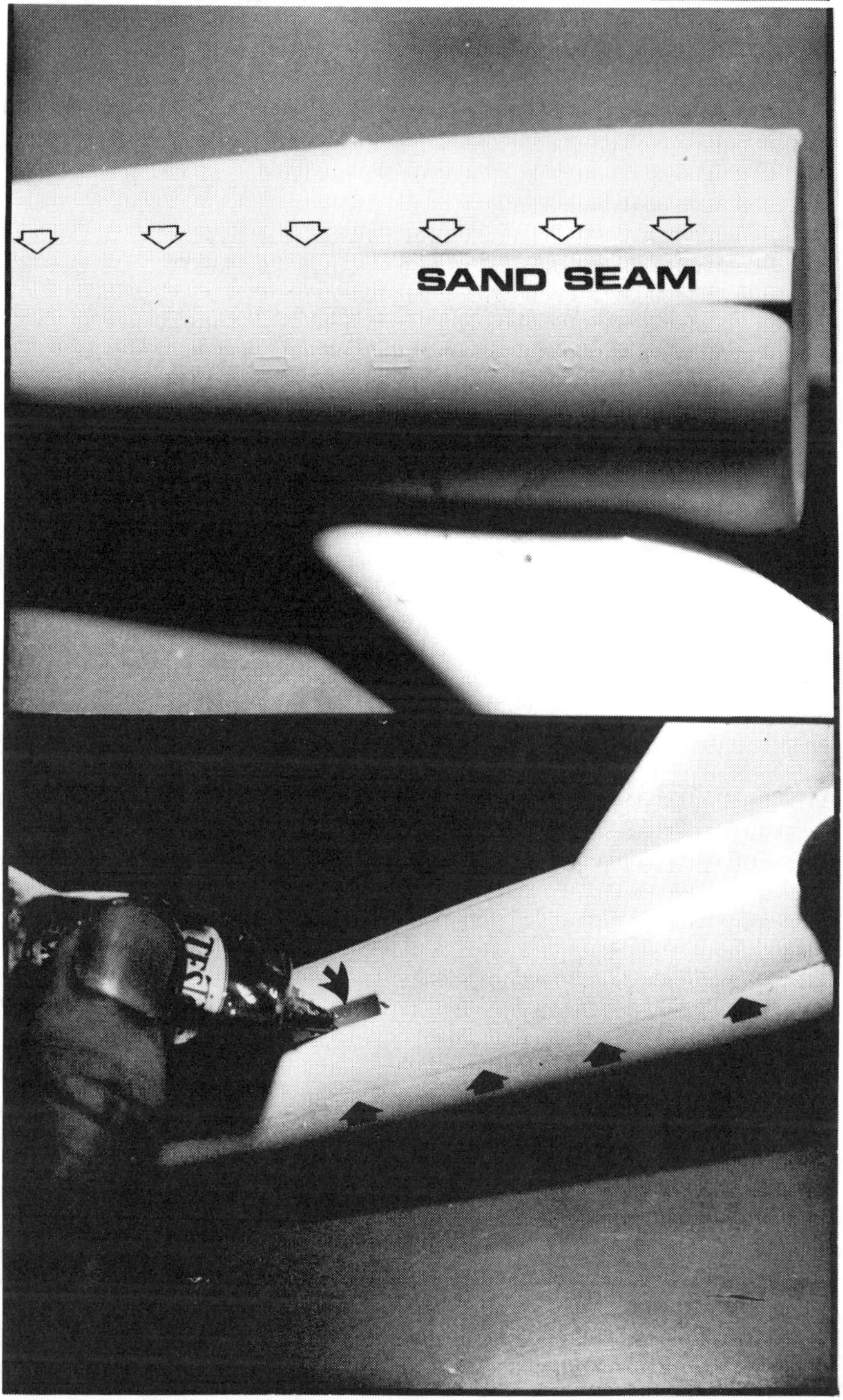
SAND SEAM

11.- AFTER THE ENTIRE SECONDARY HULL IS COMPLETED, YOU CAN START ATTACHING THE NACELLE SECTIONS. THIS PART OF ASSEMBLY WAS ONCE THE MOST DIFFICULT, BUT AMT HAS MADE IT MUCH EASIER. REMEMBER TO PLACE THE NACELLE SECTIONS SO THAT THE "SCREENS" ARE FACING EACH OTHER ON THE INSIDE. BEGIN CEMENTING* BY PLACING ONLY 2 - 3 DROPS OF CEMENT INTO THE "HOLES" OR "INSERTS", THEN INSERTING THE NACELLE SECTION. YOU WILL NOTICE A SLIGHT ANGLE AT THE BOTTOM OF NACELLE STRUT (PYLON) TO ELIMINATE "SAGGING". THE PHOTOS SHOULD SERVE AS ALIGNMENT GUIDES. THE "V" SHOULD BE A NEAR PERFECT 90° ANGLE. BY ONLY USING A SMALL AMOUNT OF CEMENT, YOU SHOULD NOT HAVE ANY EXCESS OOZING OUT WHEN YOU INSERT THE NACELLE SECTIONS. IF ANY CEMENT DOES OOZE OUT, WIPE IT AWAY QUICKLY AND CAREFULLY WITH A PAPER TOWEL. NOTICE THAT THERE IS A GAP BETWEEN THE INSERTS INSIDE THE SECONDARY HULL. THIS WILL KEEP THE NACELLES "UP" IN A 90° ANGLE. LAY THE SHIP ON ITS SIDE AND ALLOW IT TO DRY FOR A FEW HOURS.

*CRAZY GLUE CAN BE USED HERE

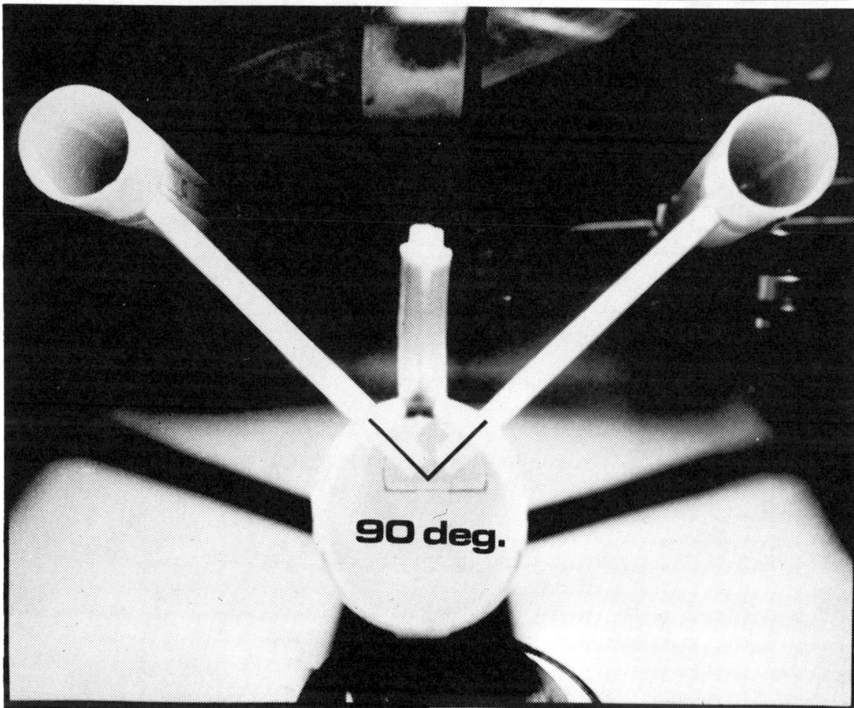
90 deg.

PRIMARY HULL

12. THE THIRD MAIN SECTION TO ASSEMBLE IS THE PRIMARY HULL. FIRST, CHECK TO SEE THAT ALL EXCESS PLASTIC AND ROUGH EDGES ARE SANDED SMOOTH. TEST-FIT BOTH HALVES TO MAKE SURE THAT THERE ARE NO GAPS. BE SURE TO FIND THE GUIDE PIN AT THE REAR OF THE HULL. WHEN YOU ARE READY TO CEMENT, APPLY A THIN AMOUNT TO THE INSIDE TOP EDGE (THE FLAT PIECE) AND PRESS BOTH HALVES TOGETHER SO THAT CONTACT IS MADE WITH THE CEMENT. ONCE CONTACT IS MADE, APPLY EXTREME PRESSURE (WITHOUT BREAKING IT) AND ADD STRIPS OF MASKING TAPE TO MAKE A TIGHT FIT. START APPLYING PRESSURE FROM THE FRONT, WORKING AROUND THE SIDES TO THE REAR. RUBBER BANDS SHOULD ALSO BE ADDED. LET DRY FOR AT LEAST 6 HOURS. LATER, WHEN YOU REMOVE THE TAPE AND RUBBER BANDS, SOME GAPS MAY RE-APPEAR. THIS CAN BE CORRECTED BY BRUSHING SOME LIQUID PLASTIC CEMENT* INTO THE GAPS AND APPLYING FINGER PRESSURE FOR ABOUT 30 - 40 SEC.

*CRAZY GLUE CAN BE USED HERE

SPREAD GLUE EVENLY ON EDGE

13. ONCE THE PRIMARY HULL HAS BEEN COMPLETED, SAND THE OUTSIDE EDGE TO GET RID OF THE LIP OR OVERHANG THAT CREATES A BAD LOOKING SEAM. SOME TRIMMING WITH A SHARP KNIFE MAY BE REQUIRED. AT THIS TIME, ALSO CEMENT IMPULSE ENGINES (THRUSTERS) TO THE HULL. AFTER THIS IS DONE, YOU CAN ATTACH IT TO THE REST OF THE SHIP. APPLY A SMALL AMOUNT OF CEMENT TO THE TOP OF THE DORSAL AND CONNECT THE PRIMARY HULL TO THE SECONDARY HULL. LAY THE SHIP ON ITS "BACK" AND ALLOW IT TO DRY. IT SHOULD LIE PERFECTLY FLAT IF ASSEMBLED CORRECTLY (AS SHOWN).

DETAILS

14. AFTER A FEW HOURS YOU CAN START ADDING OTHER DETAILS. THE FRONT OF THE SECONDARY HULL (END CAP) CAN BE ATTACHED BY APPLYING CEMENT TO THE HULL SECTION. NEXT, AFTER THAT HAS SET, THE MAIN SENSOR (DISH ANTENNA) CAN BE INSERTED. IT IS ADVISABLE THAT FOR REALISM, ABOUT 1/8" OR MORE BE TRIMMED OFF THE REAR OF THE MAIN SENSOR, SO THAT IT FITS CLOSER TO THE HULL (AS SHOWN).

15. (OPTIONAL) IF YOU WISH TO PLACE YOUR SHIP ON A STAND AS SHOWN ON AP:01:47, START BY DRILLING A HOLE INTO THE LOWER PART OF THE SECONDARY HULL. IT SHOULD BE AT THE ANGLE SHOWN, ABOUT 45° TO 50° FROM THE HORIZONTAL. THE DIAMETER OF THE HOLE DEPENDS UPON WHAT SIZE OF PLASTIC ROD YOU CHOOSE, HOWEVER, 3/8" IS A GOOD SIZE. THESE RODS ARE MADE OF STYRENE PLASTIC AND ARE REASONABLY STURDY (PLEASE REFER TO AP:01:08). YOU WILL ALSO NEED A BRASS OR COPPER ROD TO FIT SNUGLY AROUND THE PLASTIC ROD (THAT WILL BE ATTACHED TO A WOODEN BASE LATER). WHEN THE HOLE HAS BEEN DRILLED, MAKE SURE THAT THE ROD, WHEN INSERTED, MAKES A SNUG AND TIGHT FIT. PUSH IT ALL THE WAY INTO THE SHIP UNTIL IT TOUCHES THE UNDERSIDE OF THE TOP. APPLY CEMENT EVENLY WHERE THE HOLE AND ROD MEET (AS SHOWN) AND LET DRY. TOTAL LENGTH OF ROD SHOULD BE 4 - 6".

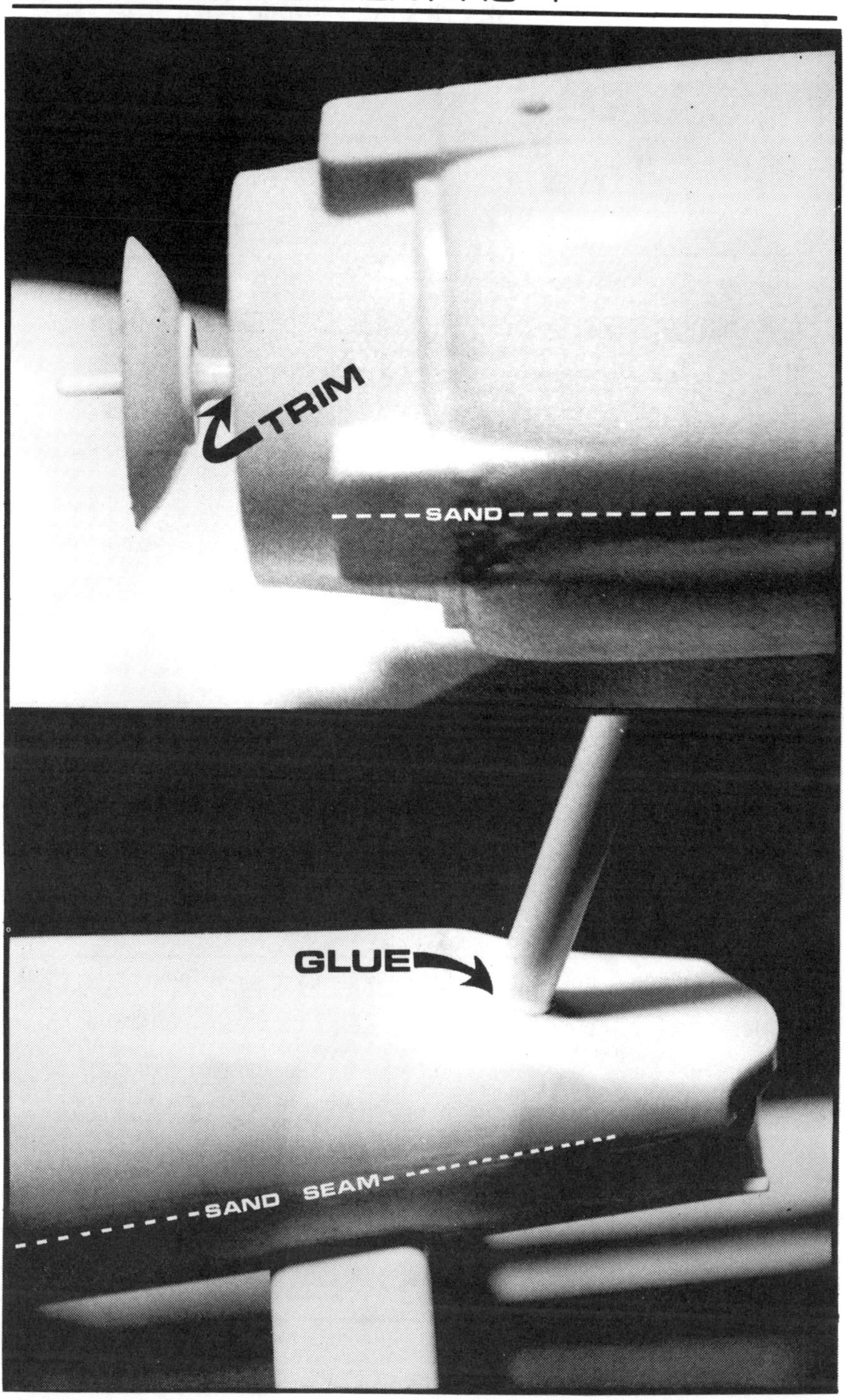
TRIM
SAND
GLUE
SAND SEAM

16. (OPTIONAL) ANOTHER MODIFICATION WHICH YOU CAN ADD IS CUTTING OUT THE SCREENS FROM THE INTERCOOLERS. THIS IS THE WAY THE INTERCOOLERS APPEARED ON THE 14 FT. STUDIO MODEL. THIS IS EASILY DONE BY MAKING 4 CUTS INTO THE SCREENS WITH A RAZOR SAW, THEN "CHUNKING" OUT THE SAWED PIECES WITH A PAIR OF NEEDLE-NOSE PLIERS. IT IS IMPORTANT TO ONLY SAW ABOUT HALF-WAY INTO THE COOLER. WHEN ALL OF THE PIECES HAVE BEEN REMOVED, SOME LIGHT FILING WILL ELIMINATE THE ROUGH SPOTS. THE MODIFIED INTERCOOLER SHOULD LOOK AS THE CENTER PHOTO SHOWS.

17. THE END CAPS OF THE NACELLES CAN NOW BE ADDED. GUIDE PINS ARE PROVIDED TO MAKE CERTAIN THAT THE CAPS ARE "CENTERED" ON THE ENDS. ONE IS INTENDED FOR THE LEFT AND ONE FOR THE RIGHT END.

18. THE DOMES FOR THE FRONT OF THE NACELLES (POWER UNIT DOMES) CAN BE ADDED BY APPLYING CEMENT TO THE DOME AND POSITIONING IT ONTO THE NACELLE UNTIL THE CEMENT HAS SET (ABOUT 30 - 40 SEC). THE POWER UNIT SHIELD IS CEMENTED DIRECTLY ON TOP OF THE FRONT END OF THE NACELLE WHEN THE SHIP IS "ON ITS BACK".* ALLOW ALL DETAILS ON THE NACELLES TO DRY FOR A FEW HOURS. (THE POWER UNIT SHIELD CAN BE ADDED NOW OR AFTER PAINTING).

*SEE STEP NO. 21.

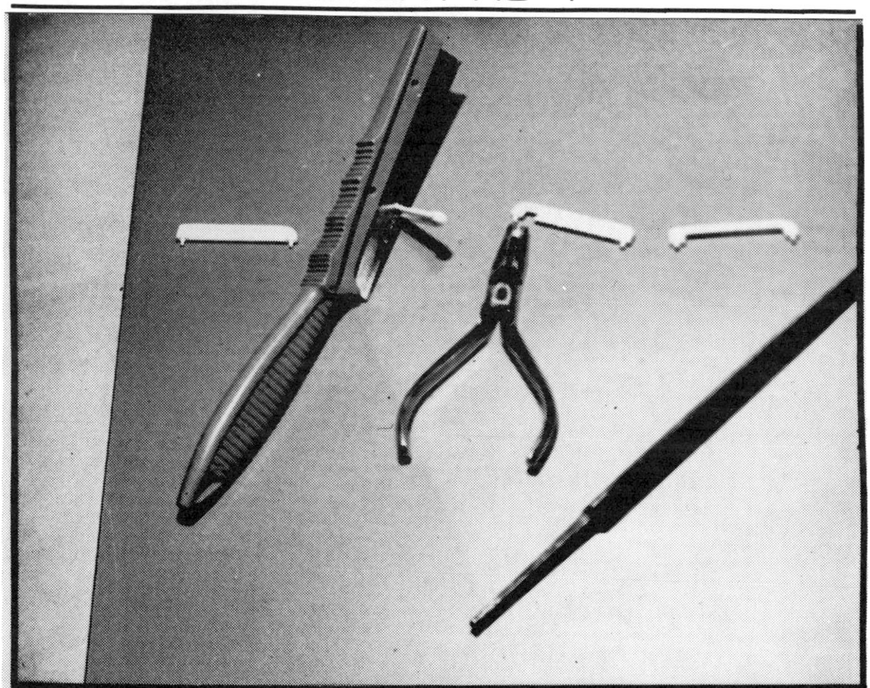

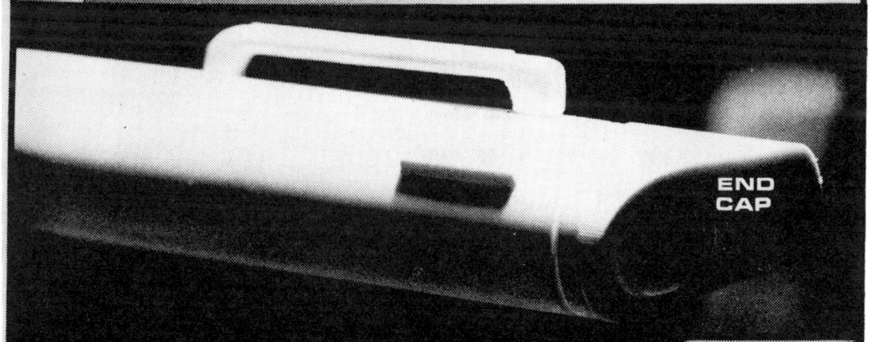
END
CAP

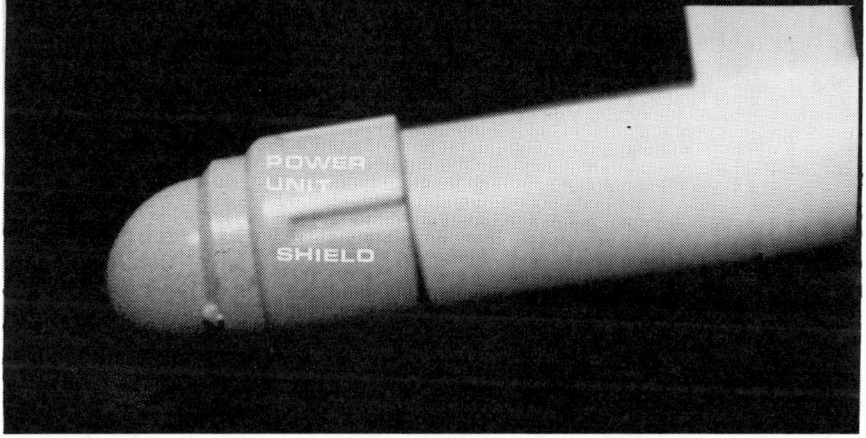
POWER
UNIT
SHIELD

PAINTING HINTS

19. AT THIS POINT YOU SHOULD DECIDE WHETHER TO PAINT YOUR STARSHIP OR TO LEAVE IT WHITE. SAND ALL SEAMS ON THE ENTIRE SHIP UNTIL THEY ARE BARELY VISIBLE. ALSO, USE SOME CONTOUR PUTTY TO FILL IN THE CRACKS AND GAPS. THE MOST OBVIOUS GAPS WILL APPEAR AT THE FRONT SIDE OF THE SECONDARY HULL AND THE SEAM AROUND THE EDGE OF THE PRIMARY HULL. WHEN YOU HAVE FINISHED THE SANDING AND FEEL THAT THE SHIP IS READY FOR PAINTING OR FINISHING, THERE ARE THREE WAYS TO ACCOMPLISH THIS:

1. LEAVE THE SHIP WHITE AFTER SANDING, APPLY DECALS, THEN SPRAY SHIP WITH A FLAT LAQUER SPRAY.
2. LEAVE THE SHIP WHITE BUT PAINT ALL DETAILS, I.E. SCREENS ON NACELLES AND STRUTS, NACELLE ENDS, FRONT DOMES AND MAIN SENSOR (SEE PAINTING SCHEMATIC), THEN APPLY DECALS.
3. PAINT ENTIRE SHIP ONE BASIC COLOR (GREY, BLUE-GREY, OR A VARIATION OF WHITE), THEN PAINT DETAILS AS SHOWN ON THE PAINTING SCHEMATIC, AND APPLY DECALS.

SHOWN IN THE UPPER RIGHT PHOTO IS AN AIRBRUSH. THIS PARTICULAR MODEL IS MADE BY BADGER AIRBRUSH CO. AND SELLS FOR ABOUT $10-$15. BY USING AN AIRBRUSH THE MODELER CAN OBTAIN FINER RESULTS THAN BY AEROSOL SPRAYING. THE PAINT GOES ON MUCH THINNER AND PRODUCES A FINER, MORE REALISTIC COAT. IF YOU PLAN TO ASSEMBLE MANY MODELS, AN INVESTMENT IN AN AIRBRUSH IS WORTHWHILE. THE SHIP SHOWN ON AP:01:49 WAS AIRBRUSHED. IF YOU DESIRE TO USE AN AEROSOL PAINT, REMEMBER TO SHAKE THE CAN VIGOROUSLY FOR ABOUT 90 SECONDS.

FLATS
AERO BLUE

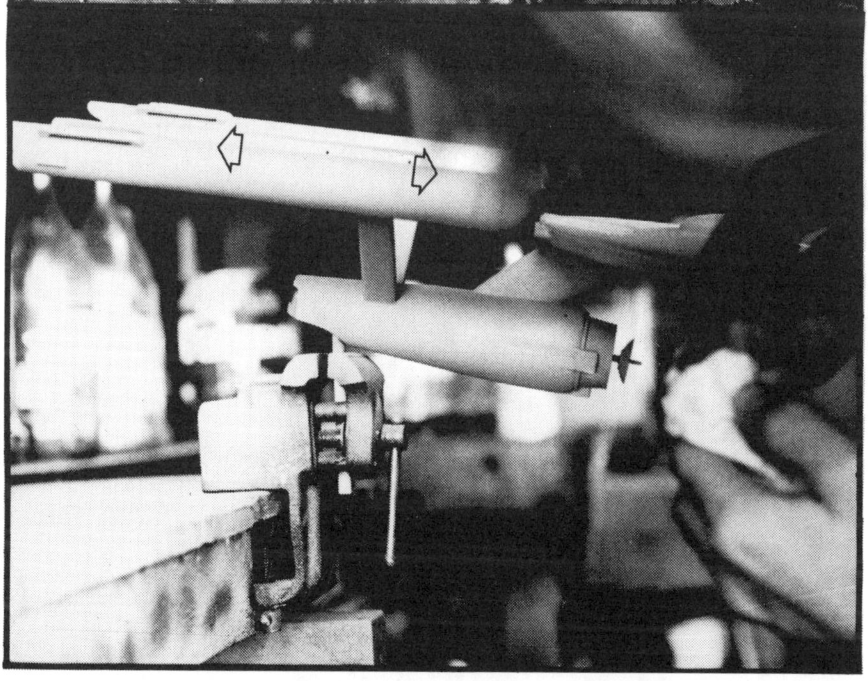

(19 cont.)

WHEN SPRAYING, HOLD THE CAN 8 - 10" AWAY, AVOIDING EXCESS PAINT THAT FORMS RUNS. USING A BACK AND FORTH MOTION, YOU WILL GIVE AN EVEN COAT THAT WON'T RUN. PAINT THE TOP FIRST, THE SIDES, AND THEN THE BOTTOM. TWO COATS MAY BE NECESSARY. SUPPORT FOR THE SHIP CAN BE PROVIDED BY THE ROD INSERTED (STEP 15) OR BY HANGING THE SHIP FROM A STRING. WHEN THE SHIP HAS BEEN PAINTED THE BASIC COLOR, ALLOW IT TO DRY OVERNIGHT. THE PAINT MAY APPEAR DRY BY THE FLAT TEXTURE OF IT, BUT ACTUALLY IT IS NOT DRY. AFTER SUFFICIENT TIME, PROCEED TO STEP 20. LISTED BELOW ARE SOME COLOR SCHEMES FOR USE BY BEGINNERS. INTERMEDIATE MODELERS MAY CONSULT PRIMER 2 FOR SOME OTHER COLORS.

BEGINNER:

BODY: PACTRA AERO BLUE OR CAMOUFLAGE GREY (LT. GREY)

DETAILS: FLAT RED, FLAT YELLOW, PACTRA STEEL (SILVER), FLAT BLACK, COPPER, FLAT WHITE
(OPTIONAL: GOLD LUMINESCENT NAIL POLISH FOR NACELLE DOMES AS A GLOSSY GLITTER-SPECIAL EFFECT).

INTERMEDIATE:

BODY: (USING AN AIRBRUSH)

MIX:

FLOQUIL PRIMER	7 PARTS
REEFER WHITE	2 PARTS
PLATINUM MIST	3 PARTS
FLOQUIL GLAZE	1 PART (OPTIONAL FOR SEMI-GLOSS)
GN BIG SKY BLUE	2 PARTS (OPTIONAL FOR BLUISH CAST)

DETAILS: SAME AS ABOVE

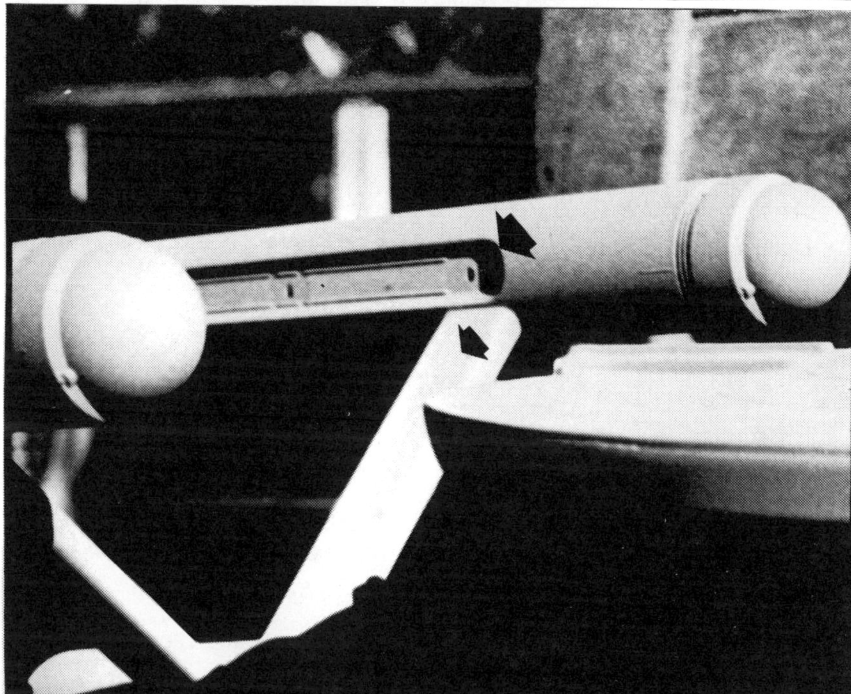

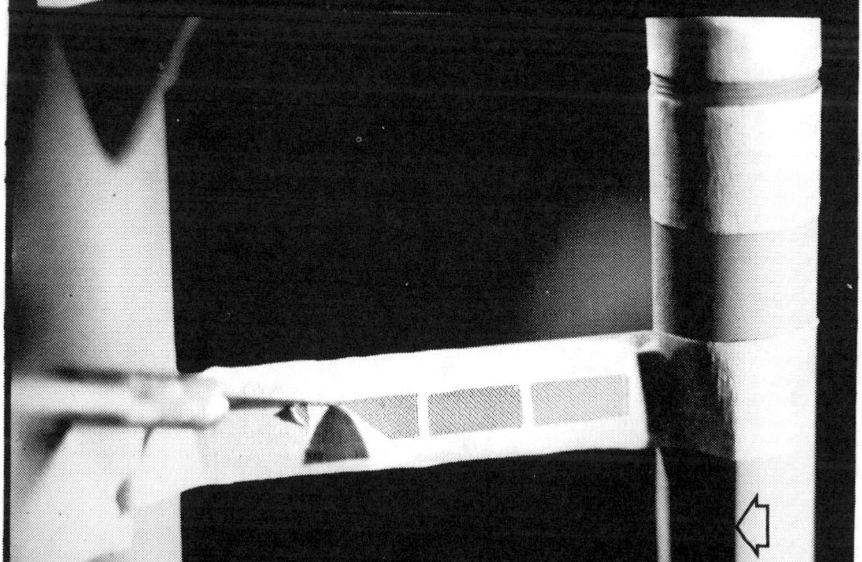

(refer to previous page)

20. WHETHER YOU CHOOSE AN OVERALL COLOR FOR YOUR STARSHIP OR PREFER TO LEAVE IT WHITE, YOU MAY WANT TO PAINT THE SCREENS AND THE FRONT END OF THE NACELLES. THIS CAN BE EASILY DONE BY FIRST PLACING MASKING TAPE OVER THE SCREENS, THEN TAKING A PENCIL OR PEN AND RUBBING AROUND THE RELIEF AREA (SCREENS). NEXT, TAKE A SHARP KNIFE AND CUT AROUND THE RELIEF AREA, AS SHOWN. REPEAT USING THE SAME METHOD ON THE SCREENS RUNNING THE LENGTH OF THE PODS (SEE ARROWS).

21. PAINT OVER THE MASKED AREA, AS SHOWN. AFTER PAINT HAS DRIED, REMOVE THE TAPE, AND THE POWER UNIT SHIELD CAN BE ADDED. IT SHOULD BE THE SAME COLOR AS THE REST OF THE SHIP.

22. THE RESULT SHOULD LOOK SIMILAR TO THIS. EXCESS BLACK PAINT CAN BE CAREFULLY SCRAPED OFF WITH A SHARP, POINTED KNIFE. AVOID SCRAPING OFF THE PAINT UNDERNEATH.

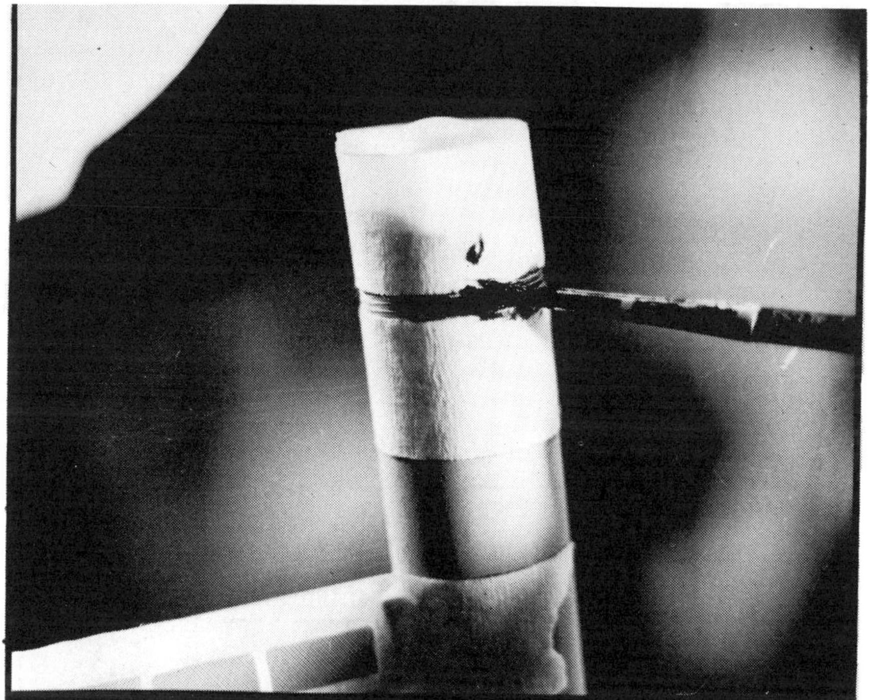

23. THIS IS WHAT A COMPLETED NACELLE POD SHOULD LOOK LIKE. IT IS LESS THE "POWER UNIT SHIELD" ON THE BOTTOM FRONT END, AND THE "CONTROL REACTORS" ON THE FACING SIDE.

24. THIS IS AN EXTREME CLOSE-UP OF THE NACELLE END. THE WHITE "BALL" IS PART OF THE AMT REVISION OF EARLY 1976. THE ACTUAL SIZE OF THE "BALLS" ARE MUCH LARGER THAN THIS AND MAY BE MODIFIED. AT THIS CLOSE RANGE OF DETAIL, ACCURACY IN MODELING IS CRITICAL AND MINOR MISTAKES IN PAINTING CAN BECOME NOTICEABLE.

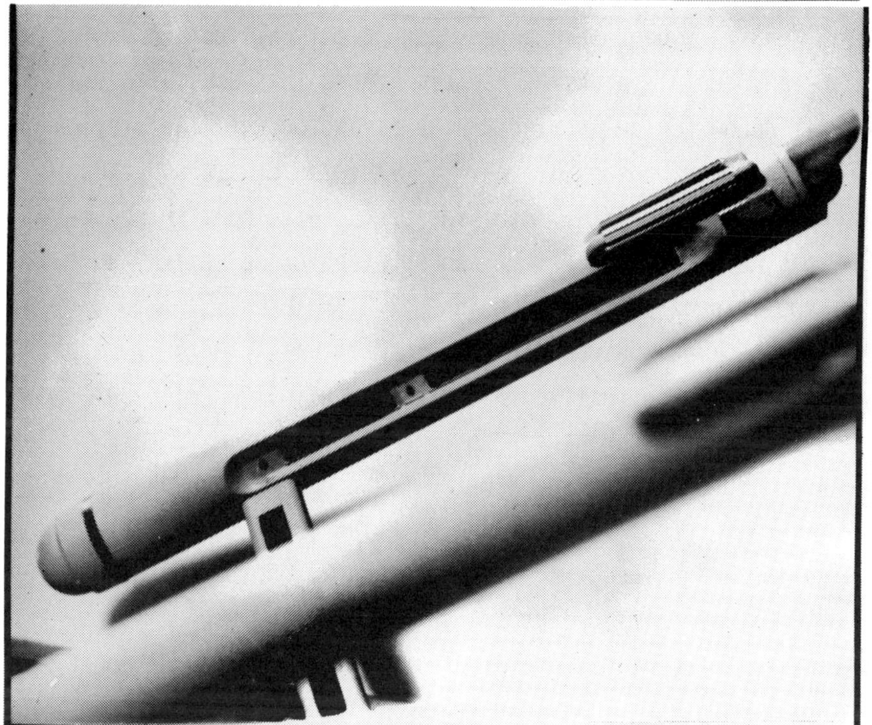

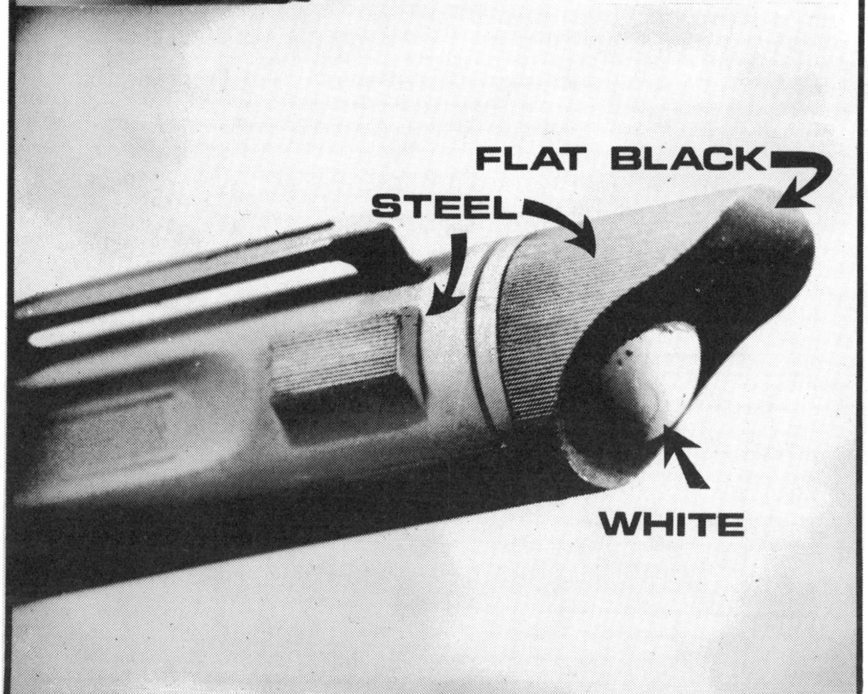
FLAT BLACK
STEEL
WHITE

25. AFTER YOUR SHIP HAS BEEN COMPLETELY PAINTED AND HAS DRIED, BEGIN ADDING THE DECALS. THOSE THAT COME WITH THE KIT ALLOW FOR THE NAMING AND NUMBERING OF ANY CONSTITUTION CLASS HEAVY CRUISER STARSHIP. STARSHIPS CAN BE NUMBERED NCC-1700 - NCC-1714.* CUT THE DECALS WITH SOME SHARP SCISSORS AS CLOSE AS POSSIBLE AROUND EVERY LARGE LETTER AND CUT GROUPINGS OF LETTERS SUCH AS "U.S.S. ENTERPRISE" INTO STRIPS. THEN, BEFORE PLACING THEM IN WATER, MATCH THEM UP TO THE SHIP SO THAT YOU ARE FAMILIAR WITH THE PLACEMENT. THE DECALS SHOULD BE PLACED IN SOME LUKEWARM WATER FOR ABOUT 30 SECONDS. THE NCC-1701 SHOULD BE PLACED ON THE PRIMARY HULL AS SHOWN ON THE PAINTING SCHEMATIC. THE POSITIONING OF THE OTHER DECALS ARE SHOWN ON THE PHOTO SUPPLEMENT OPPOSITE. CONSULT OTHER PHOTOS FOR PROPER ALIGNMENT AS WELL AS THE ONE APPEARING ON THE TOP OF THE MODEL BOX. WHEN THE DECALS HAVE BEEN POSITIONED PROPERLY, REMOVE THE EXCESS WATER WITH A DAMP PAPER TOWEL AND ALLOW THEM TO DRY. YOU MAY WANT TO ADD SOME CHAMP DECAL SET BEFORE THE DECALS COMPLETELY DRY TO ADD PERMANENCE.

*IN CHOOSING A PROPER STARSHIP NAME AND NCC NUMBER, PLEASE CHECK THE "STAR TREK CONCORDANCE", UNDER "STARSHIPS" AND THE "STARFLEET TECHNICAL MANUAL", UNDER "HEAVY CRUISER STARSHIPS". THESE TWO REFERENCES HAVE DIFFERENT NCC NUMBERS FOR ASSIGNED STARSHIP NAMES.

SPECIAL NOTE: REFER TO "TERRAN BASED REFERENCES" AP:01:50 FOR MORE DECAL INFO.

26. IF YOU CHOOSE TO ADD A BASE, AS SHOWN, STAIN THE WOOD A COLOR OF YOUR CHOICE, AND PAINT THE COPPER SUPPORT ROD FLAT BLACK. THE PLASTIC ROD THAT YOU ADDED IN STEP 15 WILL GO INTO THE COPPER ROD. SPRAY THE SHIP WITH A FLAT LAQUER SUCH AS TESTOR'S DULLCOTE OR A PHOTO MATTE SPRAY, TO PROTECT IT.

27. YOU ARE NOW COMPLETED......FINALLY!!

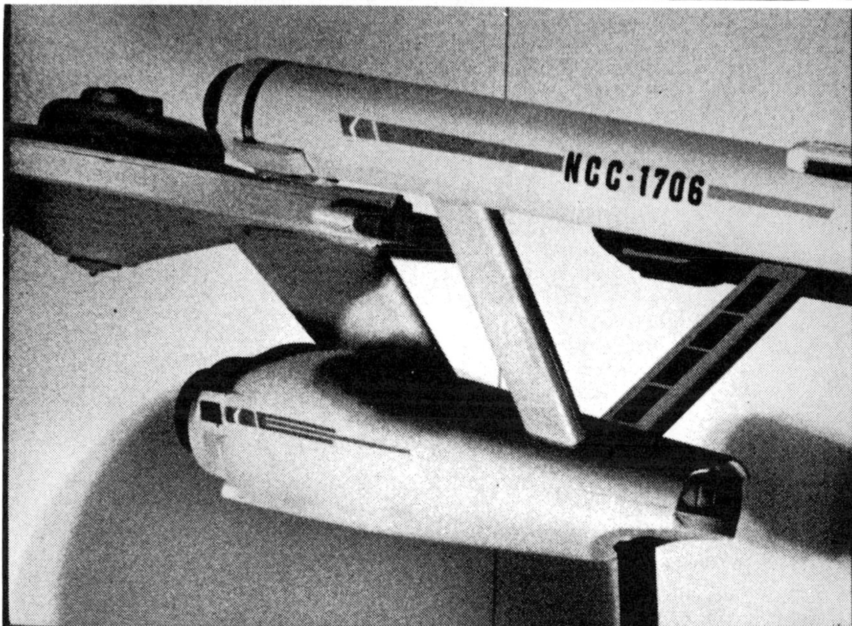
NCC-1706

NCC-1706

NCC-1706
NCC-1706
BASE

ASSEMBLY PROCEDURE: HC-1

SUPPLEMENT TO: APHC-1/pmn

SPACING OF NCC NUMERALS ON PRIMARY HULL: FACING FRONT (NOT DRAWN TO SCALE)

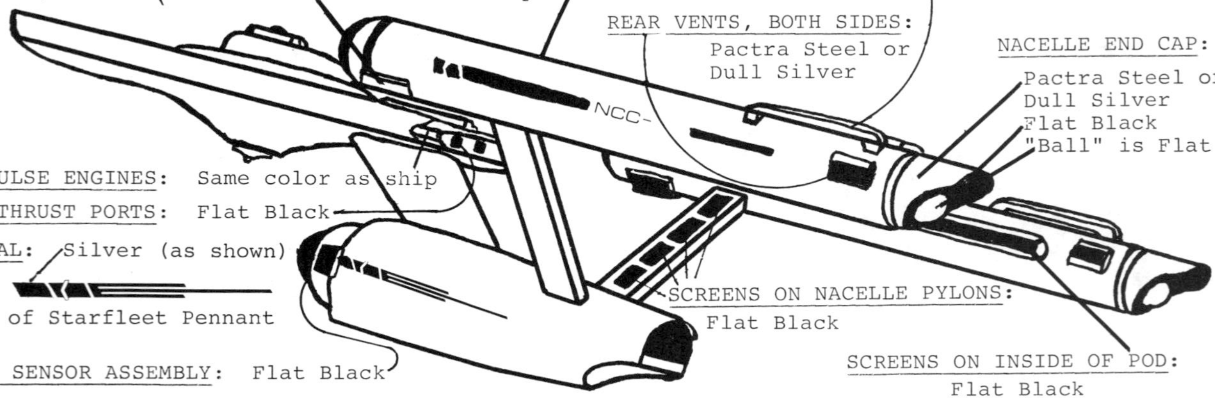

CONTROL REACTOR (not visible): Pactra Steel

ENGINE SHIELD: Same color as ship

INTERCOOLERS: Same color as ship

REAR VENTS, BOTH SIDES: Pactra Steel or Dull Silver

NACELLE END CAP: Pactra Steel or Dull Silver
Flat Black
"Ball" is Flat White

IMPULSE ENGINES: Same color as ship

THRUST PORTS: Flat Black

OPTIONAL: Silver (as shown) ahead of Starfleet Pennant

SCREENS ON NACELLE PYLONS: Flat Black

MAIN SENSOR ASSEMBLY: Flat Black

SENSOR DISH: Dull Copper

SCREENS ON INSIDE OF POD: Flat Black

NACELLE DOME:

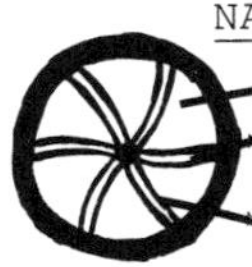

Flat Red
Flat Yellow
Cover with 1 coat of gold nail polish
Add fan blades with black felt pen

SUPPORT ROD: Flat Black

SUPPORT BASE: Stain color of your choice.

OVERALL COLOR OF HULL:

Camouflage Gray
or
Aero Blue

In aerosol spray by Pactra.

PAINTING SCHEMATIC FOR HEAVY CRUISER STARSHIPS
drawn by Mark Finlay SD 7703.13 BEGINNER'S VERSION:

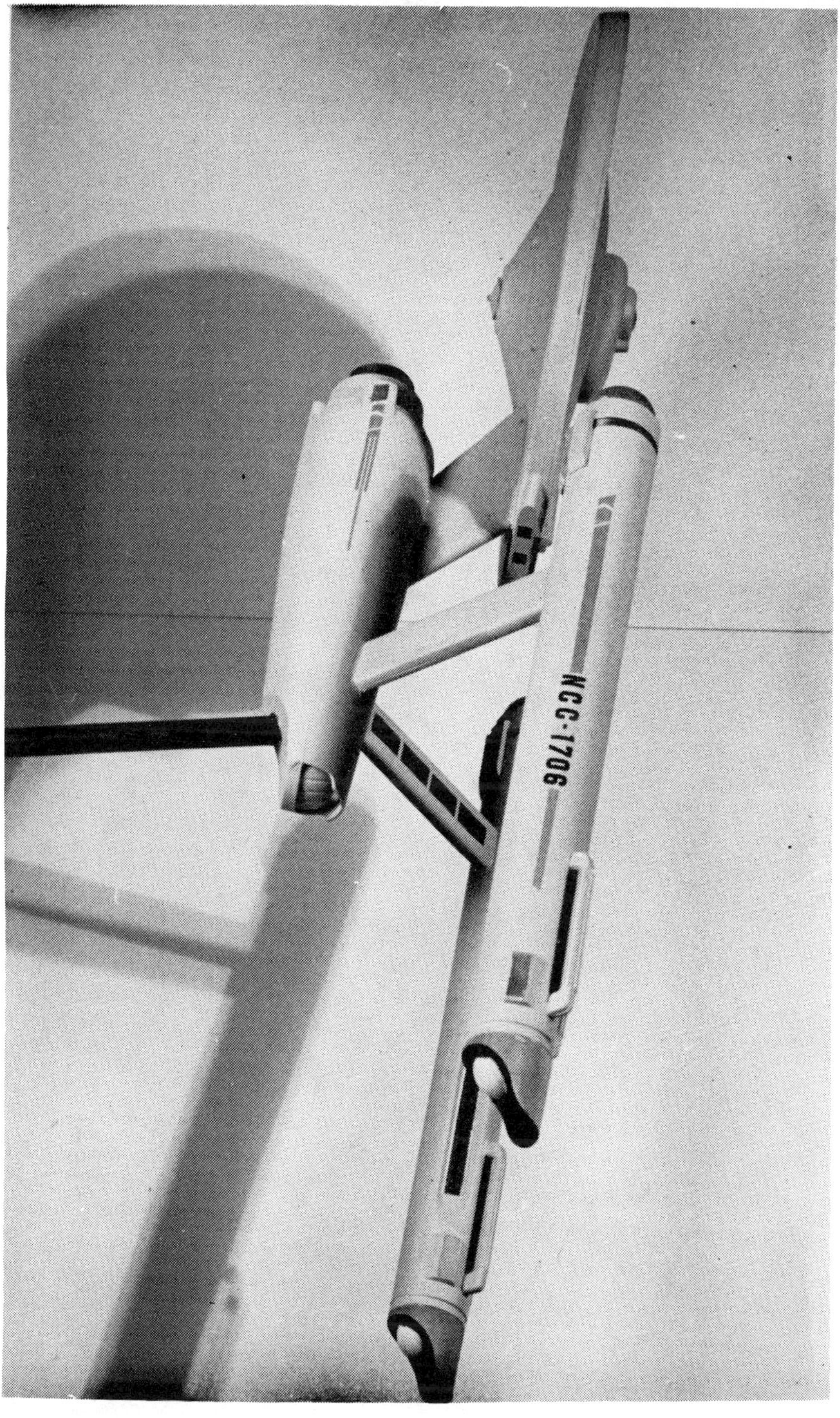
NCC-1706

1

StarFleet Assembly Manual 2

ASSEMBLY PROCEURE FOR
KLINGON BATTLECRUISER
ROMULAN BIRD·OF·PREY

a supplemental guide to assembling Klingon & Romulan spacecraft kits

conceived, written, and photographed by:
PAUL MATTHEW NEWITT

cover and centerspread artwork by:
AL SIROIS

other sketches by:
MARK FINLAY
JOHN MOORE

THIS BOOKLET IS DEDICATED TO THE SPIRIT OF IMAGINATION AND CREATIVITY

SPECIAL THANKS TO:
MOM & DAD & MARK, A.J. DESMARETZ, P.CALLANDER
STAR FLEET INTELLIGENCE, THE FANS, & JUDY BRYER

PRINTED BY A. J. DESMARETZ & CO.

first edition / first printing / JUNE 1978 M
WL 192 UCD

SPECIAL LIMITED PRINTING

SFAM-2 REFERENCE LISTING

DECLASSIFIED MATERIAL

STAR FLEET INTELLIGENCE
STAR FLEET COMMAND
UNITED FEDERATION OF PLANETS

STARDATE:3714.28SFHQ

MEMO TO: STARFLEET DIVISION
SFY/EARTH/SOL III

RE: DECLASSIFICATION/KLINGON AND ROMULAN SHIPS
MATERIAL DROPPED TO INT. LEVEL TWO

1. STAR FLEET INTELLIGENCE RECENTLY ACQUIRED A VOLUME OF INFORMATION ABOARD THE BATTLECRUISER (K) KRIEGER. WHILE ON A ROUTINE STARMAPPING MISSION, CAPT. ERIC VON RYAN OF THE USS YORKTOWN (NCC-1704) DISCOVERED THIS CRAFT, WITHOUT CREW, SEVERELY DAMAGED AND DRIFTING TOWARD STARBASE 16. ALL SYSTEMS EXCEPT THE PORT GRAF UNIT WERE IN WORKING ORDER. ALL TAPES AND RECORDS WERE CONFISCATED FOR INTERPRETATION. REASONS FOR THIS BIZARRE OCCURRENCE ARE UNKNOWN AND STILL UNDER INVESTIGATION. PART OF THE ACQUIRED MATERIALS HAVE BEEN SENT TO SFD FOR DECIPHERING AND INTERPRETATION IN YOUR DEPARTMENTS. WE HOPE THAT YOU CAN COME UP WITH SOMETHING.

2. THE ENCLOSURE INCLUDES:

   A. ASSEMBLY AND REPAIR PROCEDURES FOR KBC-2.
   B. ASSEMBLY AND REPAIR PROCEDURES FOR RPB-2.
   C. PACT WITH THE ROMULAN EMPIRE (USE OF KBC) PARTIALLY TRANSLATED.
   D. SCHEMATICS OF ROMULAN KESTREL.
   E. SCHEMATICS OF KLINGON KORVETT.
   F. COMPUTER/ARTIST ENHANCED PHOTOTRANSMISSION OF PLANET KLINGON.

3. WE SINCERELY HOPE THAT YOUR DEPARTMENTS CAN MAKE SOME USE OF THIS MATERIAL. WE BELIEVE THAT SOME OF THIS MAY BE USEFUL TO OUR PAST. SOME TWENTIETH CENTURY CIVILIANS ARE STILL HAVING PROBLEMS WITH THE PLASTIC REPRESENTATIONS OF THESE SHIPS. THE LAST TRANSMISSION TO THEM WORKED, SO THERE SHOULD BE NO REASON WHY THIS ONE SHOULDN'T EITHER. TRANSMIT ANY USEFUL INFORMATION ON SUBSPACE FREQUENCY BQ485/CE.

CODE:GT/CS68

STAR FLEET INTELLIGENCE
INTERCEPTION ACT2611

PACT

BETWEEN THE KLINGON EMPIRE AND ROMULAN EMPIRE

.....1. DUE TO THE INEFFECTIVENESS OF THE ROMULAN SHIP (BIRD OF PREY), AND DUE TO THE INVASIONS BY STARFLEET COMMAND INTO THE NEUTRAL ZONE REPEATEDLY, IT IS THE DETERMINATION OF THE ROMULAN EMPIRE TO ACQUIRE THE POSSESSION OF KLINGON DESIGNED WARSHIPS.

.....2. IN EXCHANGE FOR THE SAID NUMBER OF BATTLE-CRUISERS RECEIVED FROM THE KLINGON EMPIRE, THE ROMULAN EMPIRE WILL PROVIDE A SAID QUANTITY OF CLOAKING DEVICES TO BE INSTALLED IN VARIOUS SHIPS OF KLINGON DESIGN.

.....3. THE FIRST THREE SHIPS TO BE DELIVERED WILL BECOME THE FLAGSHIP AND SUPPORT SHIPS OF THE ROMULAN COMMANDER, TO GUARD AND DEFEND PRIME ROMULAN TERRITORY.....THESE SHIPS WILL RETAIN THE KLINGON INSIGNIA UNTIL FURTHER ORDERS. AT THAT TIME A BIRD OF PREY INSIGNIA SHALL BE APPLIED. SUCH AN INSIGNIA SHALL ADHERE TO THE SUPERIOR WARSHIP DESIGN OF THE KLINGON EMPIRE........

......THE ROMULANS SHALL OVERTAKE THE FEDERATION.......MAY THIS BE THE BEGINNING OF ROMULAN AND KLINGON ALLIED FORCES AGAINST THE FEDERATION,

..............TO THESE TERMS BOTH PARTIES AGREE.........

(PACT SIGNED BY KOMMANDER KOR, AND THE ROMULAN COMMANDER)
(DATE AND TIME UNKNOWN; END OF KNOWN TRANS-MISSION/TRANSLATION OF INTERCEPTION ACT 2611)

STAR FLEET INTELLIGENCE
INTERCEPTION ACT2611

SHIP NAME: KORVETT/SHORT RANGE FIGHTER
CULTURE: KLINGON

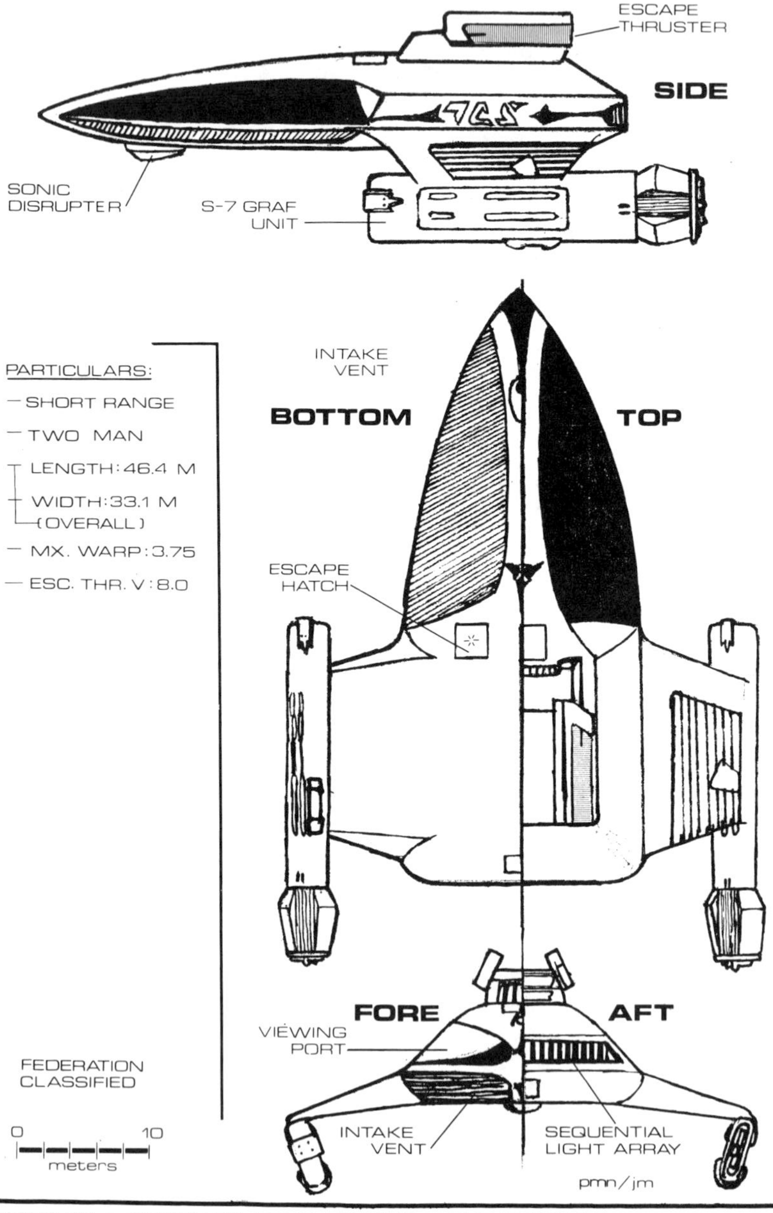

PARTICULARS:

- SHORT RANGE
- TWO MAN
- LENGTH: 46.4 M
- WIDTH: 33.1 M (OVERALL)
- MX. WARP: 3.75
- ESC. THR. V: 8.0

FEDERATION CLASSIFIED

0 10 meters

BASED ON A DESIGN BY DOUG HERRING — SUB SPACE 1,5

STAR FLEET INTELLIGENCE
INTERCEPTION / ACT2611

SHIP NAME: KESTREL / SHORT RANGE FIGHTER
CULTURE: ROMULAN

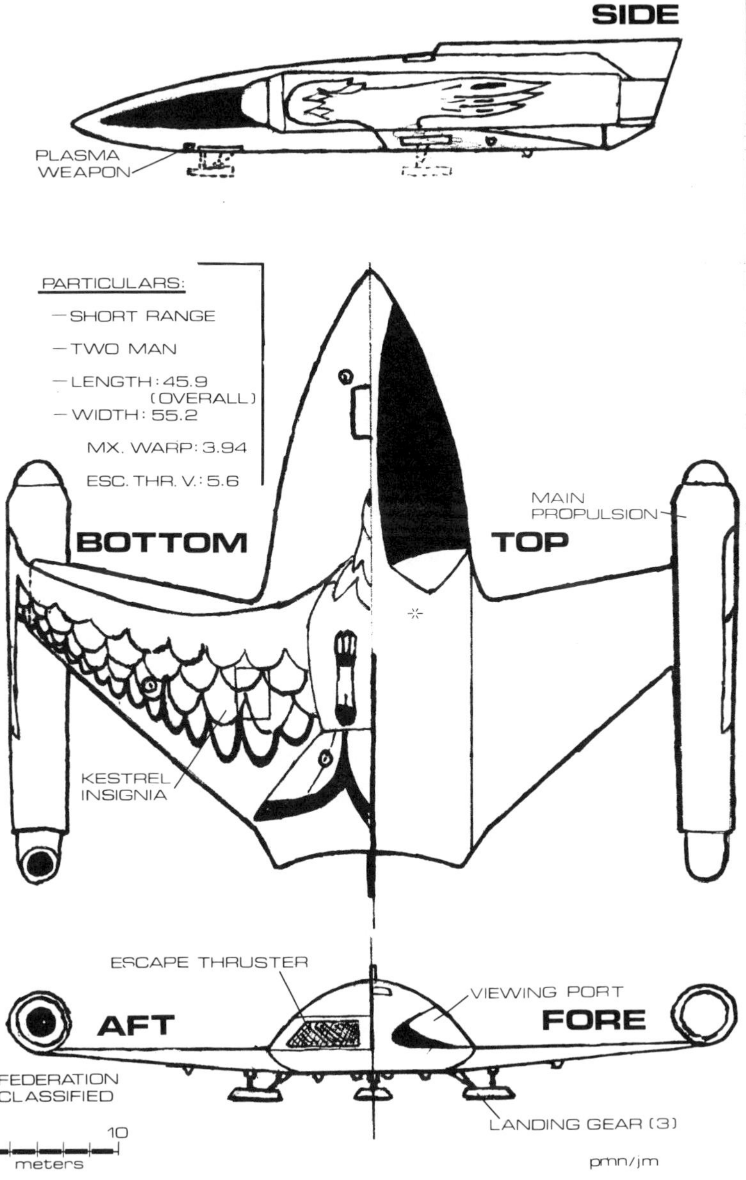

BASED ON A DESIGN BY DOUG HERRING — SUB-SPACE 1. 5

PRIMER 1 : painting and colors

KLINGON BATTLECRUISER:

The Klingon Battlecruiser's colors are of a neutral grey nature, making it difficult to obtain an exact match to the original miniature. An excellent source for obtaining these colors is the "STAR TREK GIANT POSTER BOOK #10". According to this source, the color photos show this ship to be a somewhat dark neutral grey color with a lighter variation of the same color on the "COBRA HEAD" and S-2 Graf Units (see Painting Schematic AP:02:30). This variation is more obvious in a black and white photo on page 6 in TREK #6. Both of these references are invaluable information for the discriminating modeler and are listed on page AP:01:50.

Model paint intended for military models can easily be used on this ship. Below are listed three paint manufacturers with colors suggested to match those of the original miniature:

	ENGINES AND HEAD	BODY
Humbrol	H1.5 Grey	RLM GRAU 02
Pactra	Camoflague Gray M13	Rebel Gray M12
Floquil	Primer RR9	Reefer Gray RR12

ROMULAN BIRD-OF-PREY

The Romulan Bird-of-Prey's original colors are but a partial mystery due to the scarcity of color photos of this craft. The miniature used for filming has since disappeared so much guesswork must be applied. Two color photos are available: one showing the miniature a chromate yellow color, probably a primer color, and the other photo, taken from the episode "Balance of Terror" showing only the underside. On the underside, the area (including nacelles) forward of the bird design is a dull silver color, while the area below the bird is a light grey (concrete color). The nacelle tips (rear) can be painted a reflective silver or brass color. The topside is the mystery. The rear edges are light grey. The rear fin may be light grey or dull silver, to produce a contrast. Where the two colors on the bottom meet on the topside's nacelles, is anyone's guess. You may also refer to the RBP blueprints by Mike McMaster. The painting schematics appear to be inaccurate, and suggest wing-like markings on the topside. Since there are many possibilities, please check photos and these sources and make your own decision. The color schematic on page AP:02:42 give a suggested way of painting this model. NOTE: Paint entire inside and outside flat black first if using lighting.

	UNDERSIDE BOW	UNDERSIDE STERN
Humbrol	Airframe Silver	HT.3 Surface Grey
Pactra	Grey IU3	Airframe Grey IR44
Floquil	RR100 Old Silver	Reefer White RR11

PRIMER 2: tools and materials

BOTH SHIPS:

If you choose to build these kits as simply as possible, and without lighting, here's what you'll need:

1. Fine grit sandpaper, Flexi-Grit, or emery board.
2. Plastic cement, tube or liquid, Plastiweld.
3. X-acto knife with #11 blades.
4. Paint, spray, paint brush.
5. Champ Decal Set, bowl of lukewarm water.

If you decide to add lighting, here's the added materials you'll need:

1. Plexiglas rod, 5/8" diameter, 3" long.
2. Fiberoptic: 40 mil or 56 mil diameter. DuPont Crofon Fiberoptic,
3. Light bulbs: for nacelle tips (RBP), Radio Shack Incandescent Lamps. "Leads" Lamps, 1.5 volts, 6.0 volts, 12.0 volts. Voltage dependent on preference. Yellow Jumbo L.E.D.'s can be used as well.
   Miniature Flourescent Lamps: used for model RR passenger cars. Include bulbs and sockets in kit. Available at Model R.R. Shops or direct from Wm. K. Walthers, Inc.,
4. Transparent colored acetate: available as gelatin filters (photo, theatrical lamps) or available at stationery stores as colored tabs for binder page separaters.
5. To install these lights you will need:
   a. Razor saw
   b. Electric drill or Dremel Tool with 1/8" size bit
   c. Super Glue 3 or similar
   d. Scotch tape
   e. Hook-up wire (fine, insulated)
   f. Body putty (Green Squadron putty, Testor's putty)
6. If you plan to add some battle damage, you will need these items:
   a. Dremel Tool or flexible shaft tool
   b. Steel saw bit, small carving cutter bit, etc.
   c. Pieces of detailed parts from other models to simulate engine parts
   d. Foil wrappers from cupcakes (HoHo's) and straws
   e. Flat Black, Steel paint
   f. Soldering iron (cheapie)
   g. Razor Saw (optional)
   h. Extremely fine wire (piano, electrical)

Klingon BattleCruiser

The Klingon Battlecruiser is one of Star Trek's most beautiful ships. It equals the Enterprise in design quality, and it should because it was produced by Matt Jefferies, the designer of the starship Enterprise. Mr. Jefferies spent several weeks going over sketches and ideas for this alien craft, and as you can see, his time and effort paid off. The original miniature was carved out of balsa wood and chromed mylar and plastic details were added. This miniature measures 29 inches long and now resides at the Smithsonian in Washington, D.C. along with the 11 ft. Enterprise. AMT used this miniature when making the molds for this kit, so in a way, every kit is an exact duplicate of the original. The washboard on the front of the secondary hull wasn't on the original; why AMT added it is unknown. This is the most accurate kit AMT manufactures in the Star Trek series, and produces a very striking miniature. Estes Model Rocket Co. of Penrose, Colorado, also produces this miniature, in roughly the same scale, as a flying model rocket. The detail and craftmanship is equally as good as AMT. Estes conferred with experts on Star Trek miniatures to accurately produce their flying versions of the Klingon Battlecruiser as well as the U.S.S. Enterprise.

The three letters and Klingon logo were handpainted by Mr. Jefferies. These letters have spurred the creation of a Klingon alphabet. Every ship carries these symbols in the Klingon fleet. No markings to designate other names of ships or class numbers are given, thus causing problems for the Federation.

1. AFTER OPENING THE BOX, YOU SHOULD DISCOVER A PLASTIC BAG WITH PARTS MOUNTED ON SPRUES, AN INSTRUCTION SHEET, A SET OF DECALS, AND SOME CLEAR GREEN PARTS. CHECK WITH THE INSTRUCTION SHEET TO SEE IF ANY PARTS ARE MISSING.

2. MOST OF THE LARGER PARTS CAN BE REMOVED BY FLEXING THE SPRUES OFF. HOWEVER, SOME PARTS MAY REQUIRE THE USE OF AN X-ACTO KNIFE FOR EASY REMOVAL AS SHOWN.

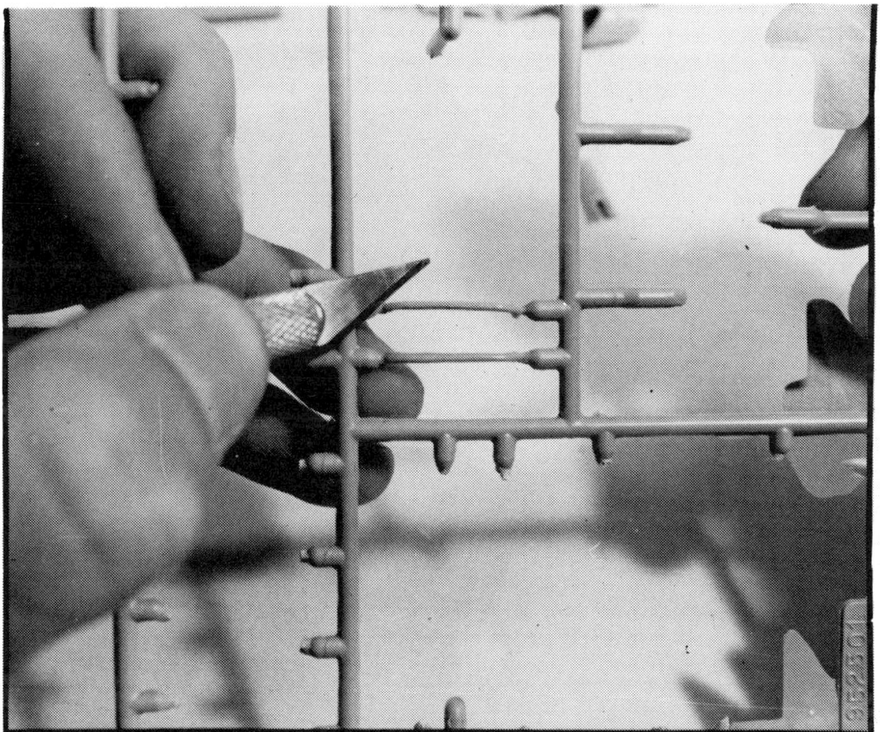

KLINGON
BATTLE CRUISER
FROM THE "KLINGON WARRIOR EMPIRE"
STAR TREK

3. GROUP THE PARTS INTO THE RESPECTIVE SECTIONS. FIRST, FIND ALL OF THE PARTS THAT BELONG TO THE "POWER PODS". THESE UNITS ARE ACTUALLY KNOWN AS S-2 GRAF UNITS, BUT FOR CONTINUITY WE'LL USE THE NAMES FROM THE INSTRUCTION SHEET. THERE SHOULD BE 6 PARTS FOR EACH POD.

4. THE TWO HALVES SHOULD GO TOGETHER WITHOUT ANY PROBLEM. USE MASKING TAPE TO SECURE THE HALVES TOGETHER. THIS WILL ELIMINATE ANY CRACKS LATER. LAY ASIDE TO DRY FOR 2 HOURS. THE POWER POD INSERTS MAY NOT BE CHROMED (SOME KITS DO NOT INCLUDE CHROME PARTS). SAND OFF THE EXCESS OR OVERHANG OF THE SEAMS FOR A MORE PROFESSIONAL LOOK AND ATTACH THE POWER POD INSERTS, POWER POD LIGHTS AND END CAPS.

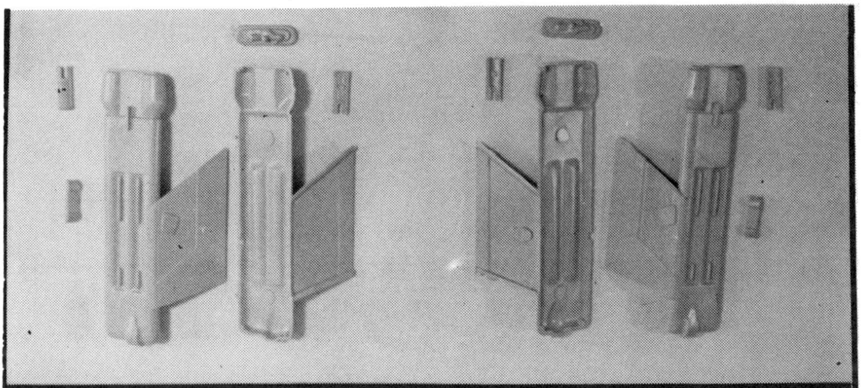

5. NEXT, WHILE THE PODS ARE DRYING, COLLECT THE PARTS TO THE "UPPER DECK". THERE SHOULD BE TWELVE PARTS. START BY GLUING THE LOWER DECK TO THE UPPER DECK. ADD THE LOWER DECK SIDES, AS SHOWN. THIS PART IS TRICKY, BECAUSE THE PARTS DO NOT FIT WELL. TRY TEST FITTING BEFORE GLUING. WHEN DRY, ADD BODY FILLER PUTTY TO THE LARGE CRACKS. WHILE THAT IS DRYING, ASSEMBLE THE "UPPER DECK SIDE PANEL BACKS" TO THE "UPPER DECK SIDE PANELS", AND AFTER BOTH HAVE DRIED, SAND OFF THE OVERHANG FROM THE PANEL BACKS. AFTER COMPLETING BOTH SETS, GLUE THEM ON EACH SIDE OF THE ASSEMBLED UPPER AND LOWER DECK PIECES, FLUSH WITH THE TOP OF THE UPPER DECK. ADD THE "LOWER DECK REAR" TO GUIDE THE ENTIRE ASSEMBLY PROPERLY. THESE PARTS MAY REQUIRE MUCH TEST FITTING BEFORE ACTUAL GLUING, SO FAMILIARIZE YOURSELF WITH THEM.

6. LASTLY, ADD THE INSERT TO THE FRONT UNDERSIDE OF THE LOWER DECK, AND THE TWO "UPPER DECK SIDE PANEL INSERTS". THESE PARTS MAY OR MAY NOT BE CHROMED. IF NOT, THEY CAN BE PAINTED LATER. IF THEY ARE CHROMED, ADD THEM AFTER THE SHIP HAS BEEN PAINTED, IF YOU SO DESIRE.

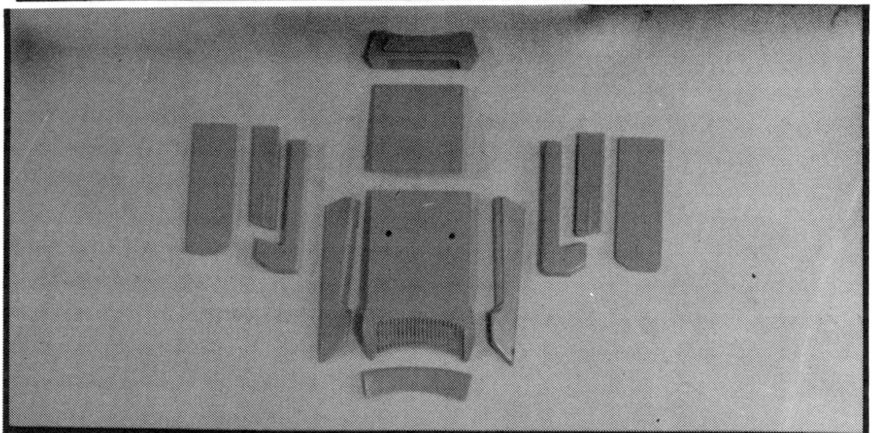

LOWER DECK SIDES

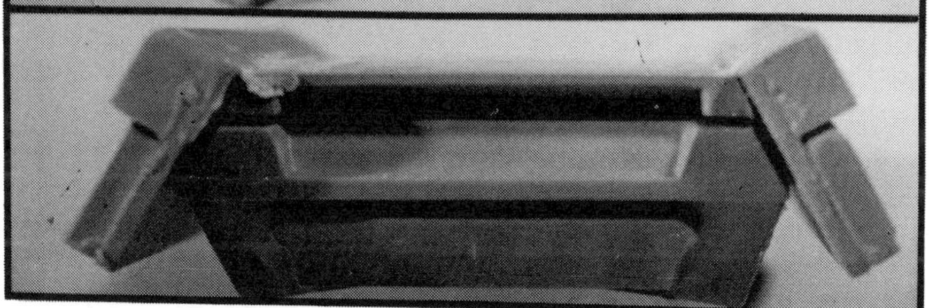

7A. THE "BOOM" ASSEMBLY IS NEXT. GATHER ALL PARTS FOR THIS SECTION. THERE SHOULD BE A TOTAL OF EIGHT PARTS. IF YOU CHOOSE TO ADD LIGHTING SEE 7B, IF NOT, CONTINUE BY FIRST CEMENTING THE TWO HALVES TOGETHER AND CAREFULLY ALIGNING THEM FOR ACCURACY. APPLY PRESSURE QUICKLY TO THE ASSEMBLED HALVES AND APPLY MASKING TAPE, THEN ALLOW THEM TO DRY. THE "BOOM SIDE INSERTS" AND "B.R. TUBES" CAN BE ADDED. THESE PARTS MAY BE CHROMED AND YOU MAY WISH TO ADD THEM LATER. REMEMBER, WHEN GLUING CHROMED PARTS TO SCRAPE THE AREA TO BE CEMENTED OFF WITH AN X-ACTO KNIFE. IF YOU'RE NOT ADDING LIGHTS, CONTINUE TO 8A.

7B. SHOWN AT RIGHT IS THE INSTALLATION OF A SMALL FLUORESCENT-LIKE BULB TO ILLUMINATE THE PRIMARY HULL. THE WINDOWS HAVE ALREADY BEEN DRILLED OUT. IF YOU CHOOSE TO COLOR THE INTERIOR LIGHTS, USE SOME TRANS-PARENT ACETATE IN LIGHT YELLOW, RED OR BLUE. TAKE SOME SCOTCH TAPE AND TAPE SMALL PIECES OF THE ACETATE TO THE INSIDE OF THE WINDOWS. NEXT, SOLDER WIRES TO THE LEADS OF THE LIGHT SOCKET AND GLUE INTO PLACE WITH EPOXY OR EQUIVALENT. ONE HALF SHOULD HAVE THE SOCKET GLUED TO IT. AFTER THAT HAS DRIED, BRING THE OTHER HALF INTO PLACE AND CEMENT, MAKING SURE THAT ABOUT ONE FOOT OF WIRE HAS RUN DOWN THE NECK OF THE BOOM AS SHOWN. ALLOW THE ASSEMBLY TO DRY. NOTE: IF YOU WANT THE GREEN SECTIONS OF THE HEAD TO ILLUMINATE, PLACE A PIECE OF FOIL DIRECTLY UNDERNEATH THE LIGHT INSIDE THE HEAD, AND ALSO CUT SOME HOLES INTO THE LOWER HALF OF THE "FOREDECK".

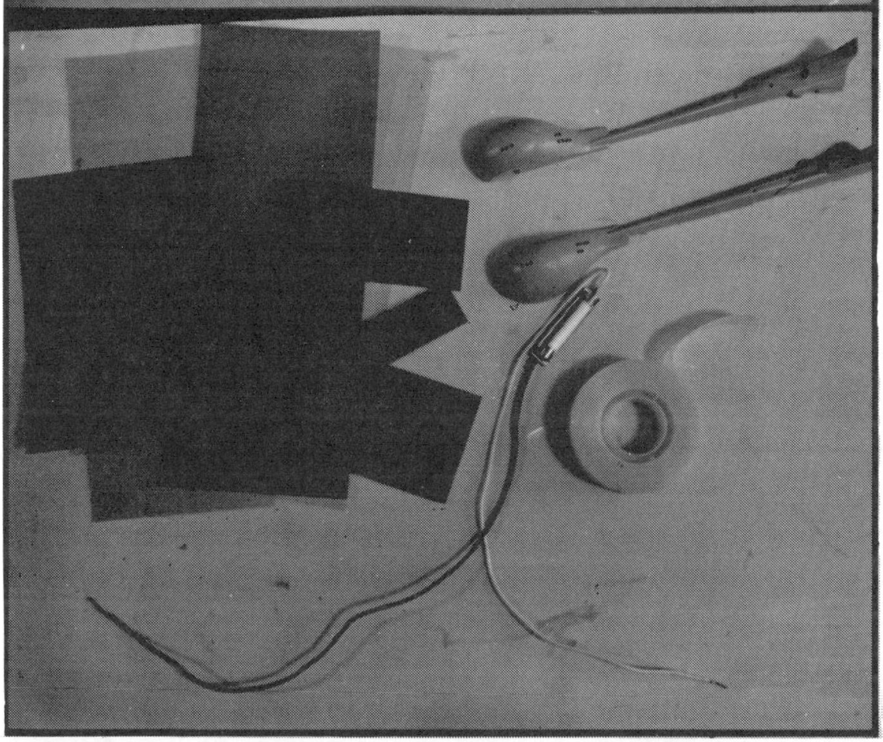

8A. THE FOREDECK, CONSISTING OF EIGHT PIECÉS, IS NEXT. START BY CEMENTING THE FOREDECK WINDOWS TO UPPER FOREDECK SECTION, THEN BRING THE LOWER FOREDECK SECTION INTO PLACE. WRAP MASKING TAPE AROUND THE EDGES TO INSURE NO CRACKS. THEN ASSEMBLE THE TWO CABIN SECTIONS IN THE SAME WAY. WHEN DRY, CAREFULLY SAND THE EDGES AND SEAMS. THE CABIN POSTS CAN BE ADDED. DO NOT ALLOW THESE TO PROTRUDE AS THE INSTRUCTION SHEET SHOWS. THE HEAD OF THESE PARTS ARE FLUSH WITH THE TOP OF THE CABIN (AS SHOWN).

8B. IF YOU WISH TO ADD FIBER OPTIC TO THE FRONT OF THE FOREDECK, DRILL OUT THE WINDOW AREAS SUGGESTED BY SMALL RECTANGLES. FOR 10-20 MIL FIBER OPTIC, USE A DRILL BIT EQUIVALENT TO A PIN IN DIAMETER. INSERT AND GLUE FIBER OPTIC INTO THE HOLES OF THE FOREDECK ALLOWING ABOUT 1/4" EXCESS STICKING OUT. THIS EXCESS WILL BE TRIMMED OFF WITH NAIL CUTTERS AFTER THE SHIP HAS BEEN PAINTED. RUN THE OTHER END OF THE FIBER OPTIC STRANDS TO THE LIGHT SOURCE. FOR BETTER BRIGHTNESS, MAKE SURE THE ENDS POINT TOWARD THE LIGHT BULB.

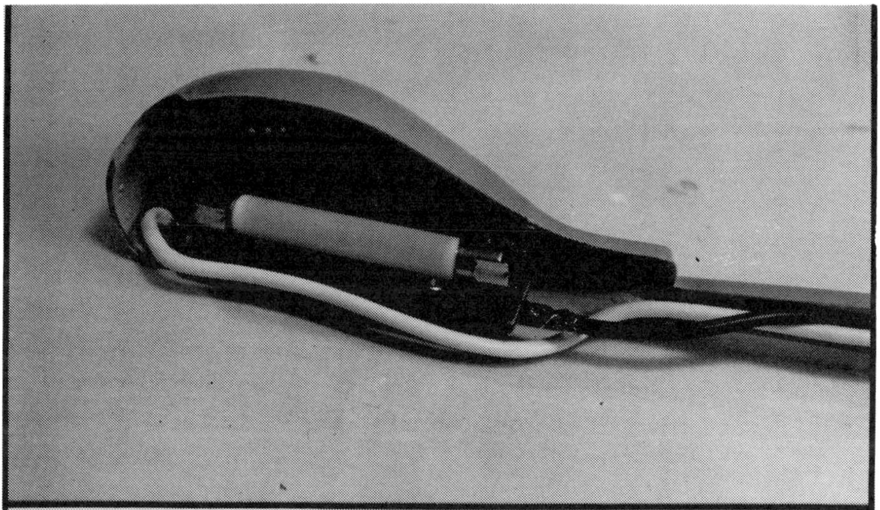

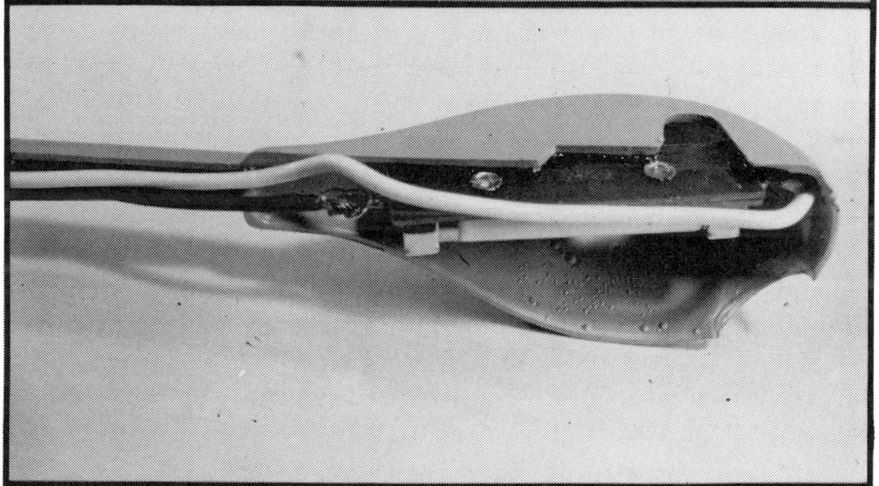

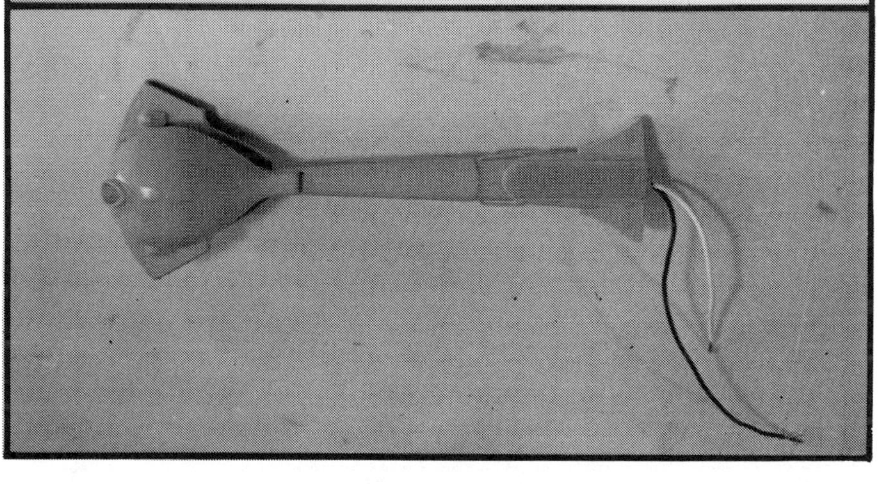

9. JOIN THE ASSEMBLED FOREDECK TO THE FRONT OF THE BOOM ASSEMBLY. THERE IS A LARGE SURFACE AREA FOR CONTACT, SO SPREAD THE GLUE EVENLY AND WHEN BOTH ASSEMBLIES MAKE CONTACT, APPLY FIRM PRESSURE TO ELIMINATE ANY CRACKS. TO MAKE SURE OF A FLAT CONTACT, SAND EACH SURFACE BEFORE GLUING. IF FIRM PRESSURE IS APPLIED, MASKING TAPE WON'T BE NECESSARY HERE. IF CRACKS ARE UNAVOIDABLE, APPLYING SOME ENAMEL NAIL POLISH WILL SUCCESSFULLY FILL IN THE CRACKS, WORKING BETTER THAN BODY FILLER PUTTY FOR SUCH A TIGHT SPACE.

10. LASTLY, INSERT THE "ANTENNA". THIS PART IS TRICKY BECAUSE IT CAN BECOME LOST INSIDE THE "HEAD" SECTION. BY USING A ROD OR DOWEL OF A DIAMETER TO FIT INSIDE THE ANTENNA, POSITIONING CAN BE SUCCESSFUL. STRONGER GLUE, LIKE "SUPER GLUE 3" OR EPOXY SHOULD BE USED, THOUGH CAREFULLY. THE ACTUAL MINIATURE HAD THE ANTENNA FRONT FLUSH WITH THE BODY SURFACE, HOWEVER, THE POSITION SHOWN HERE IS ACCEPTABLE AS WELL.

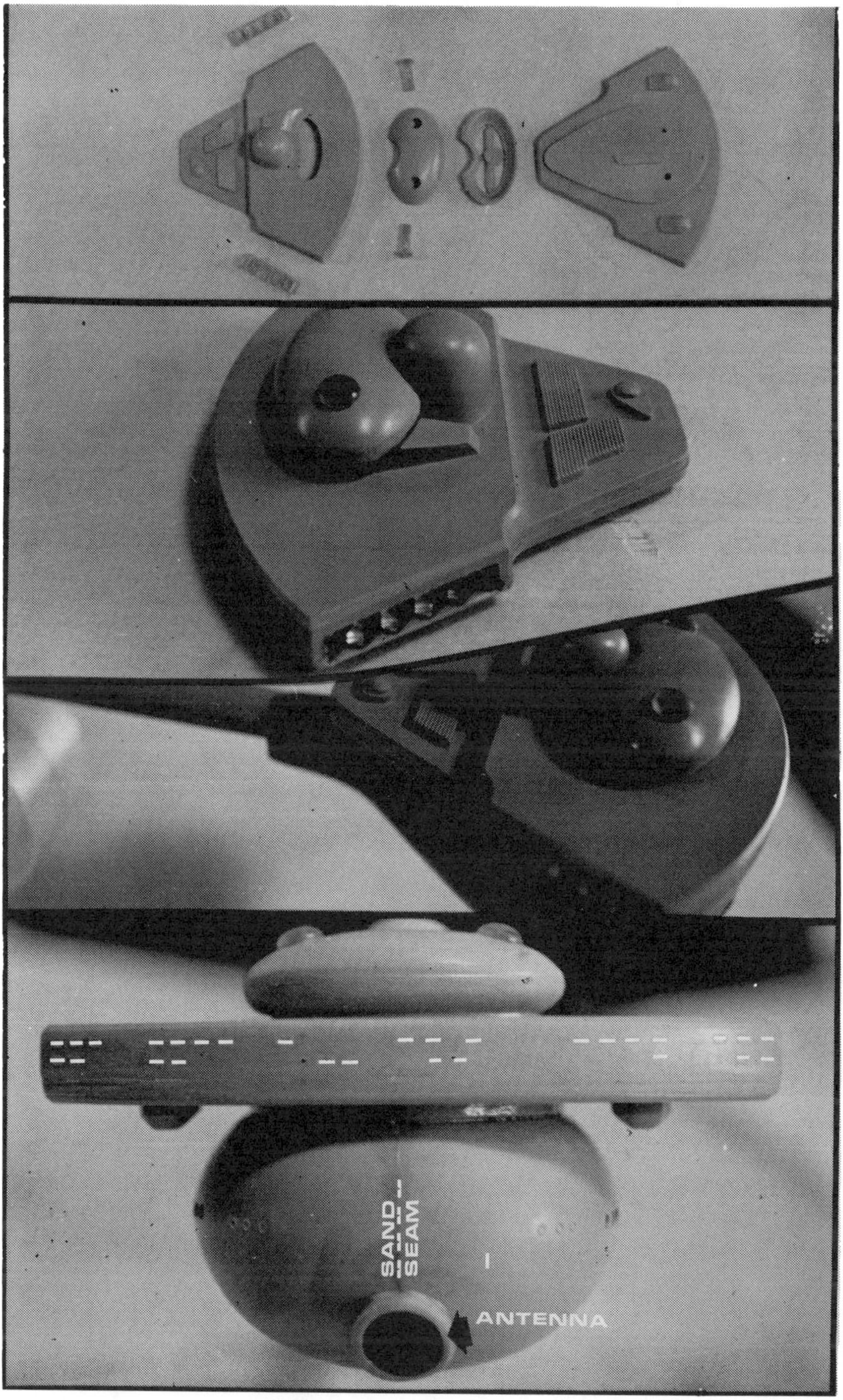
SAND SEAM
ANTENNA

11. THE SECONDARY HULL (KNOWN AS THE "HALL" ON THE INSTRUCTION SHEET) IS NEXT. THIS IS ALSO TRICKY, SO FAMILIARIZE YOURSELF WITH IT BEFORE GLUING. RUN LIQUID OR BRUSH GLUE ON ALL OF THE POINTS OF CONTACT. THE MOST IMPORTANT PLACES ARE SHOWN BY ARROWS. THE WASHBOARD SECTION IS A PARTICULAR PROBLEM. APPLYING PRESSURE IS VERY IMPORTANT WHEN GLUE IS DRYING. SHOWN AT THE TOP RIGHT ARE TWO PINS TO BE INSERTED INTO THE BOTTOM OF THE LOWER HALF. ONE IS FOR USE AS A FILL-IN PLUG, AND THE OTHER, WITH THE NOTCH, IS FOR THREAD OR FISHLINE TO BE WOUND AROUND IT, AND STRUNG UP THROUGH THE TOP HOLE IN FRONT OF THE LOWER DECK SECTION. ALSO SHOWN ARE WIRES PROTRUDING THROUGH THE HOLE WHERE THE BOOM ASSEMBLY ATTACHES. IF YOU WISH, A 9V BATTERY TERMINAL CLIP CAN BE INCLUDED HERE. THE WIRES FROM BOOM ASSEMBLY WILL ATTACH TO THESE WIRES.

12. TO INSURE AN ACCURATE FIT, CAREFULLY POSITION THE EDGES OF THE HULL HALVES WHEN GLUING THEM TOGETHER. ATTACHING MASKING TAPE TIGHTLY WILL HELP PREVENT CRACKS AS WELL. ALLOW TO DRY.

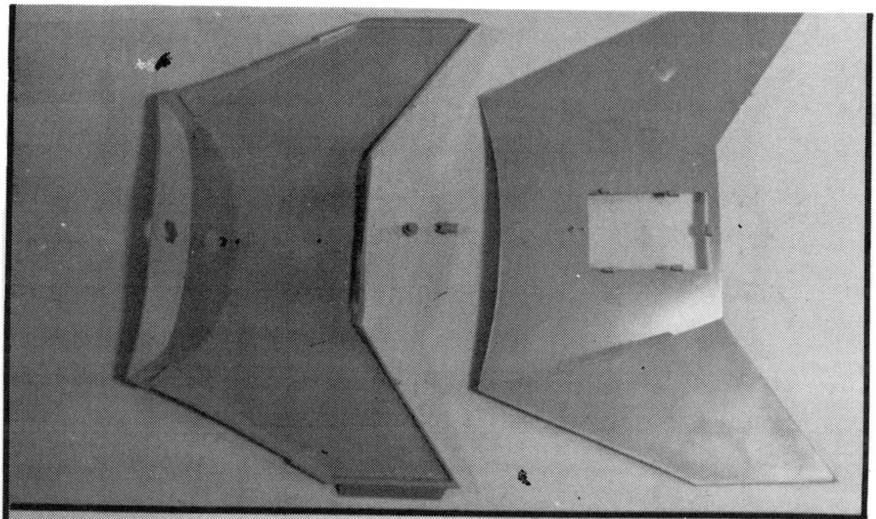

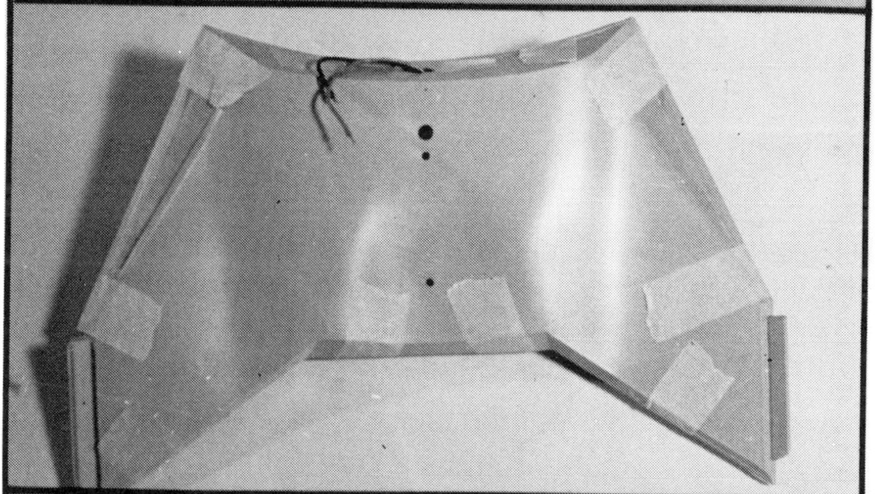

13. SHOWN AT RIGHT IS THE BOOM ASSEMBLY WITH THE "BOOM SIDE INSERTS". "B.R. TUBES" AND "LOWER BOOM INSERT" ATTACHED. THE WIRES FROM THE LIGHTING SYSTEM ARE PROTRUDING OUT THE END, AND WILL BE ATTACHED TO THE WIRES SHOWN IN THE LOWER PHOTO.

14. NOTICE THE UNEVEN SEAM OF THE HULL HALVES. MUCH SANDING WILL HAVE TO BE DONE TO ALLEVIATE THIS PROBLEM.

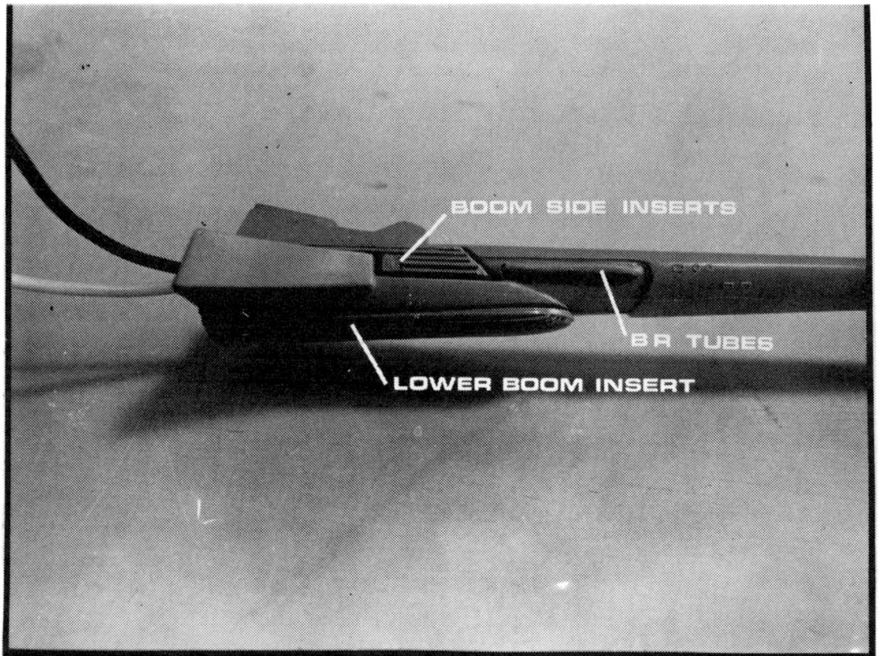
BOOM SIDE INSERTS
B R TUBES
LOWER BOOM INSERT

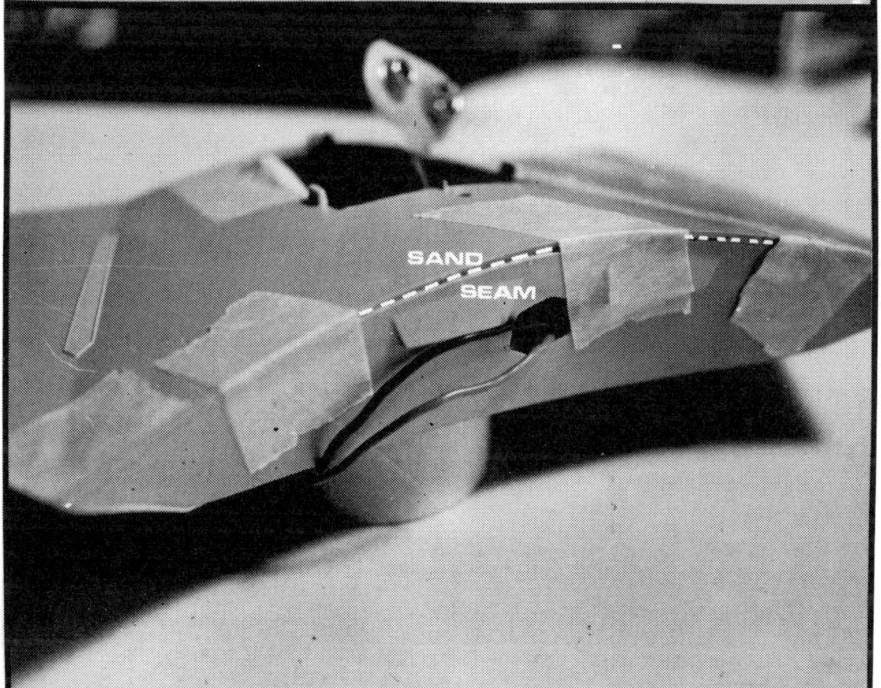
SAND
SEAM

©SIROIS 77

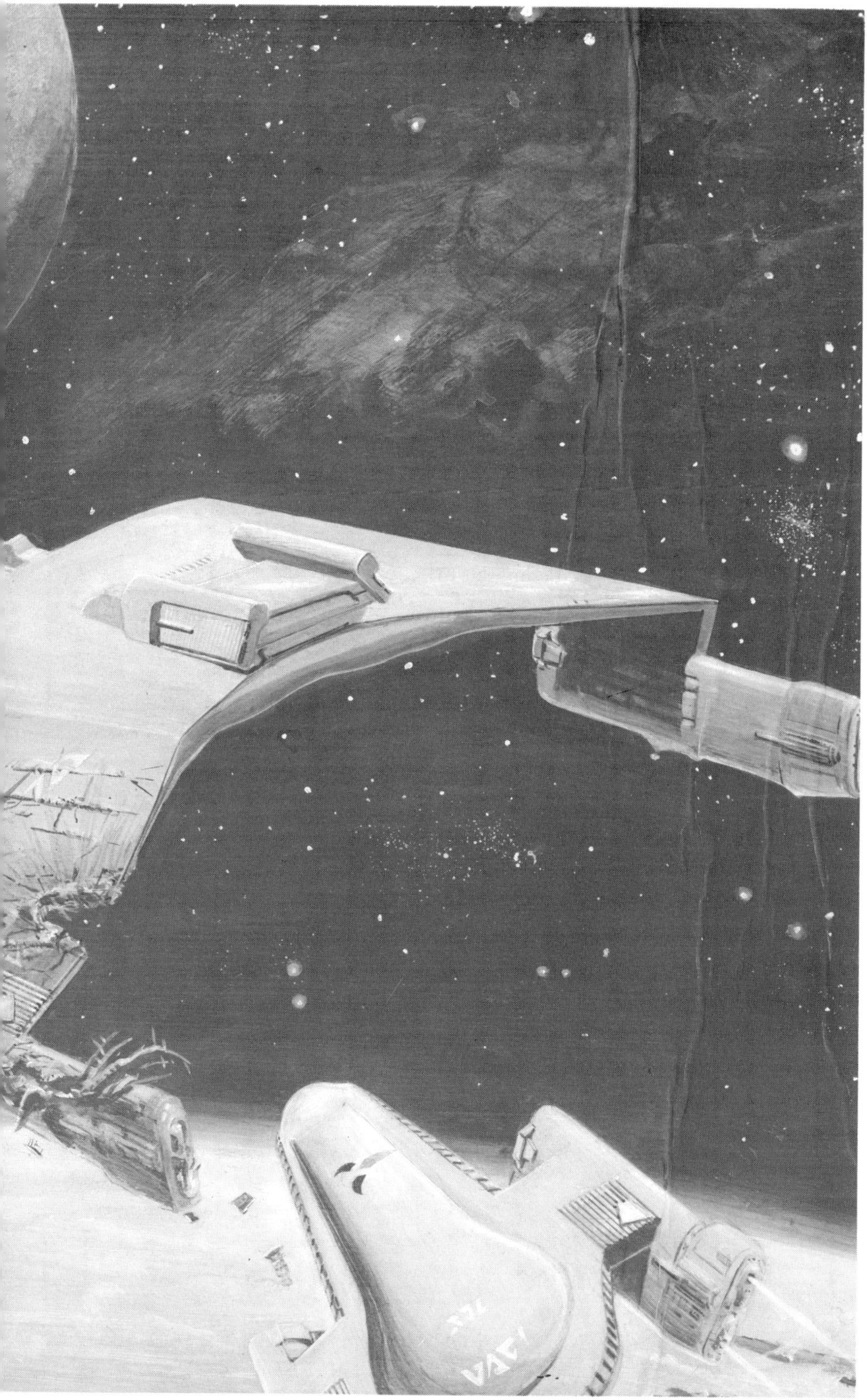

15. BOTH SECTIONS ARE SHOWN BEFORE ATTACHING THEM TOGETHER. USE A VERY STRONG GLUE FOR THIS PART. EPOXY WORKS WELL. SOME TRIMMING OF EACH PART TO MAKE CONTACT MAY BE REQUIRED. AGAIN, TEST FIT TO MAKE SURE THERE IS NO INCOMPATIBILITY.

16. NOTICE THE CRACK ON THE TOP RIGHT SECTION OF THE LOWER DECK ASSEMBLY. THIS MIGHT HAPPEN DURING PLACEMENT ON TOP OF THE HULL SECTION. EITHER ADD SOME BODY FILLER PUTTY TO THE CRACK OR RE-GLUE, APPLYING FIRM PRESSURE.

17. AT THIS STAGE ALSO ATTACH THE POWER PODS, FULLY PRE-ASSEMBLED, TO THE PROPER PLACES ON THE HULL. THERE WILL BE AN UNEVENESS ALSO THAT WILL REQUIRE SOME SANDING OFF AND SOME BODY PUTTY.

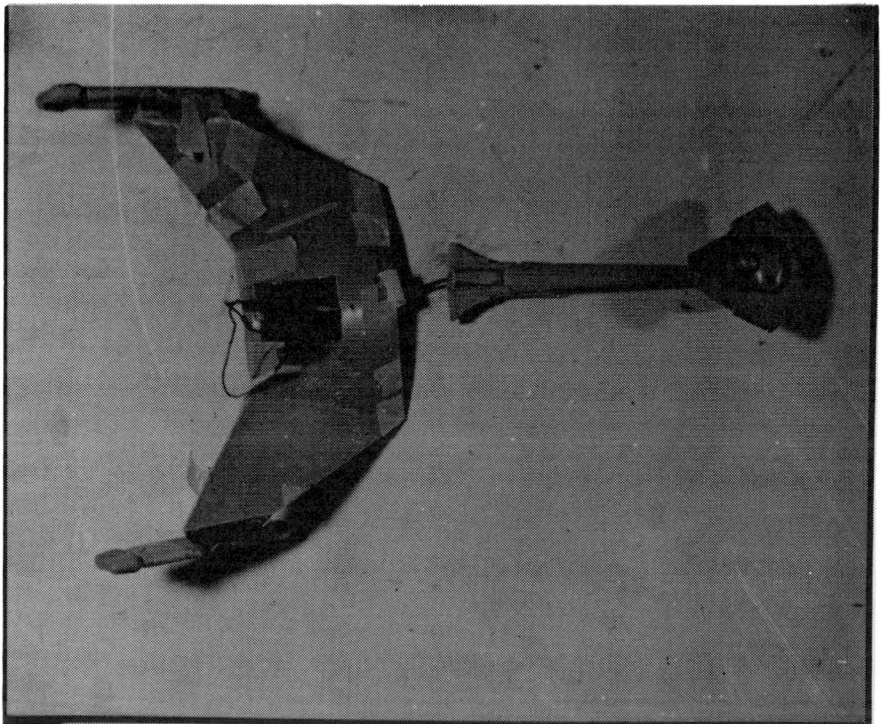

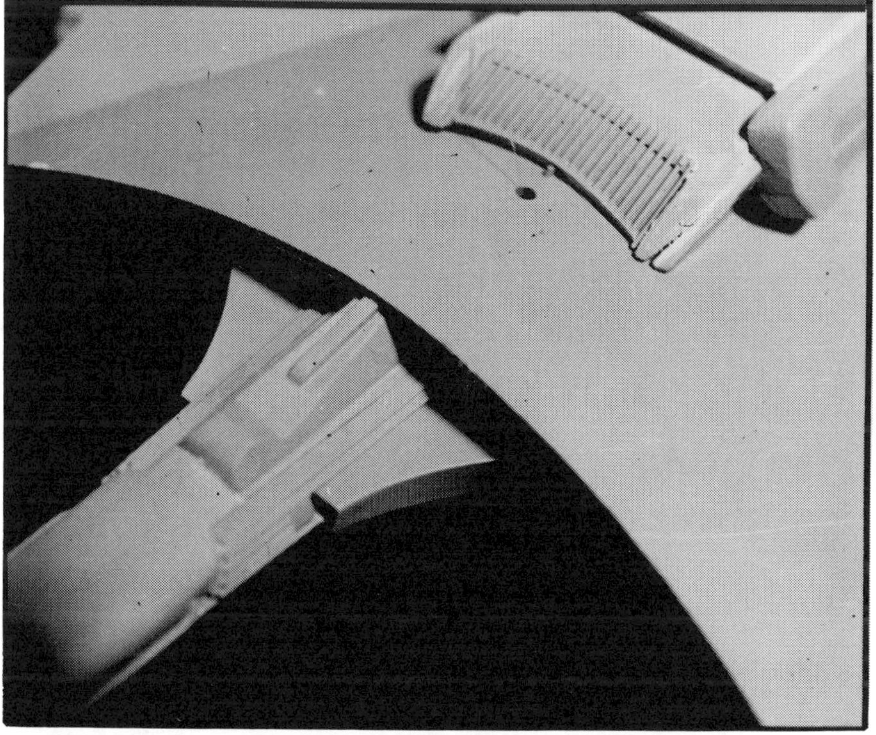

18. THESE PHOTOS SHOW THE PLACEMENT OF THE 9V BATTERY FOR THE LIGHTING SYSTEM. THE BATTERY WILL BALANCE OUT THE SHIP IF IT IS HUNG FROM A STRING. (THE OLD VERSION OF THIS MODEL KIT PROVIDED A PLACE FOR TWO "AA" BATTERIES TO ILLUMINATE A GRAIN OF WHEAT BULB IN THE HEAD. A SWITCH WAS PLACED ON THE UNDERSIDE FOR ON-OFF).

19. AFTER THE SHIP HAS BEEN PUTTIED AND SANDED TO YOUR SATISFACTION, BEGIN TO MASK OFF AREAS FOR PAINTING. SINCE THE WINDOWS ALREADY HAVE ACETATE IN THEM, THEY WILL HAVE TO BE MASKED WITH TINY PIECES OF TAPE, AS WELL AS ALL THE GREEN GLASS ON THE FOREDECK. REBEL GREY (OR EQUIVALENT) SHOULD BE APPLIED TO THE MAIN BODY, EXCEPT THE ENGINES AND HEAD (THE PAINTING SCHEMATIC WILL GIVE BETTER DETAILS ON PLACEMENT). AFTER THIS COLOR HAS DRIED, MASK OFF THE NARROWEST PART OF THE NECK AND THE AREA ABOVE THE ENGINES. PAINT THE HEAD AND ENGINES A LIGHTER GREY, SUCH AS CAMOUFLAGE GREY. REFER TO PHOTOS AND THE PAINTING SCHEMATIC FOR DETAILS.

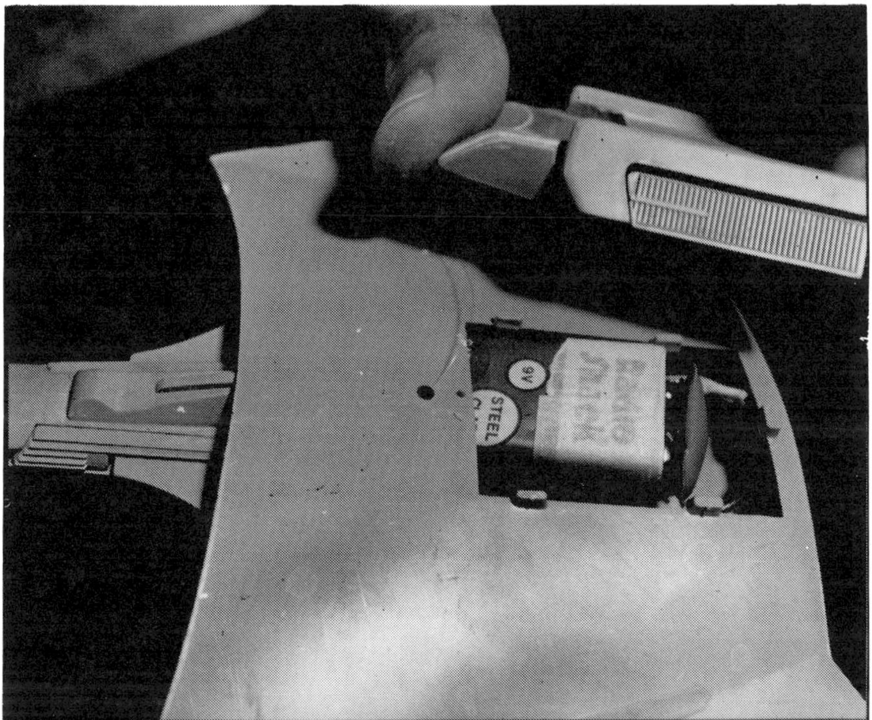
9V
STEEL

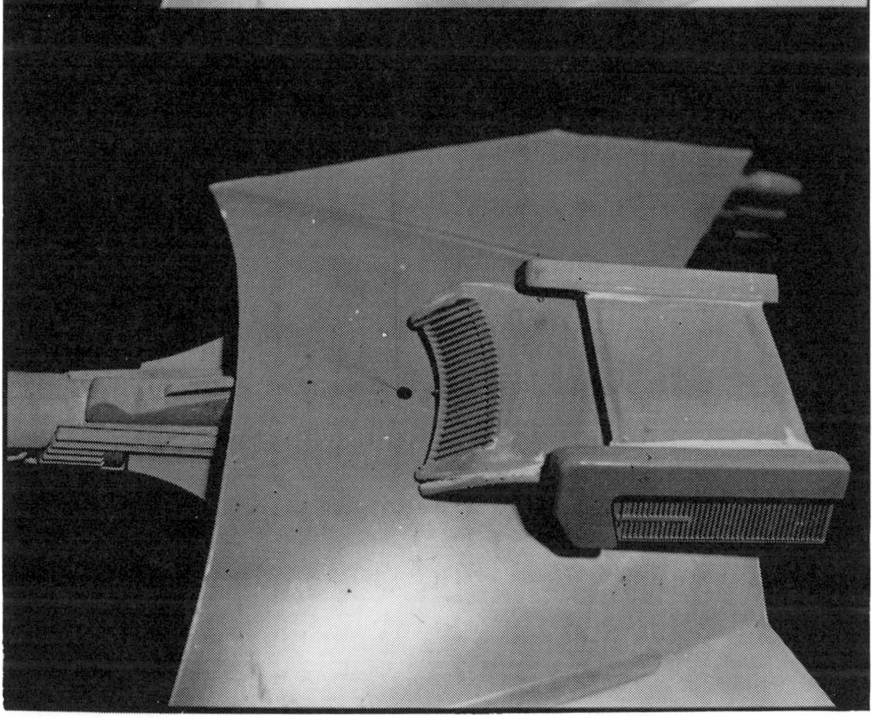

PAINTING SCHEMATIC FOR KLINGON BATTLECRUISER KBC-2

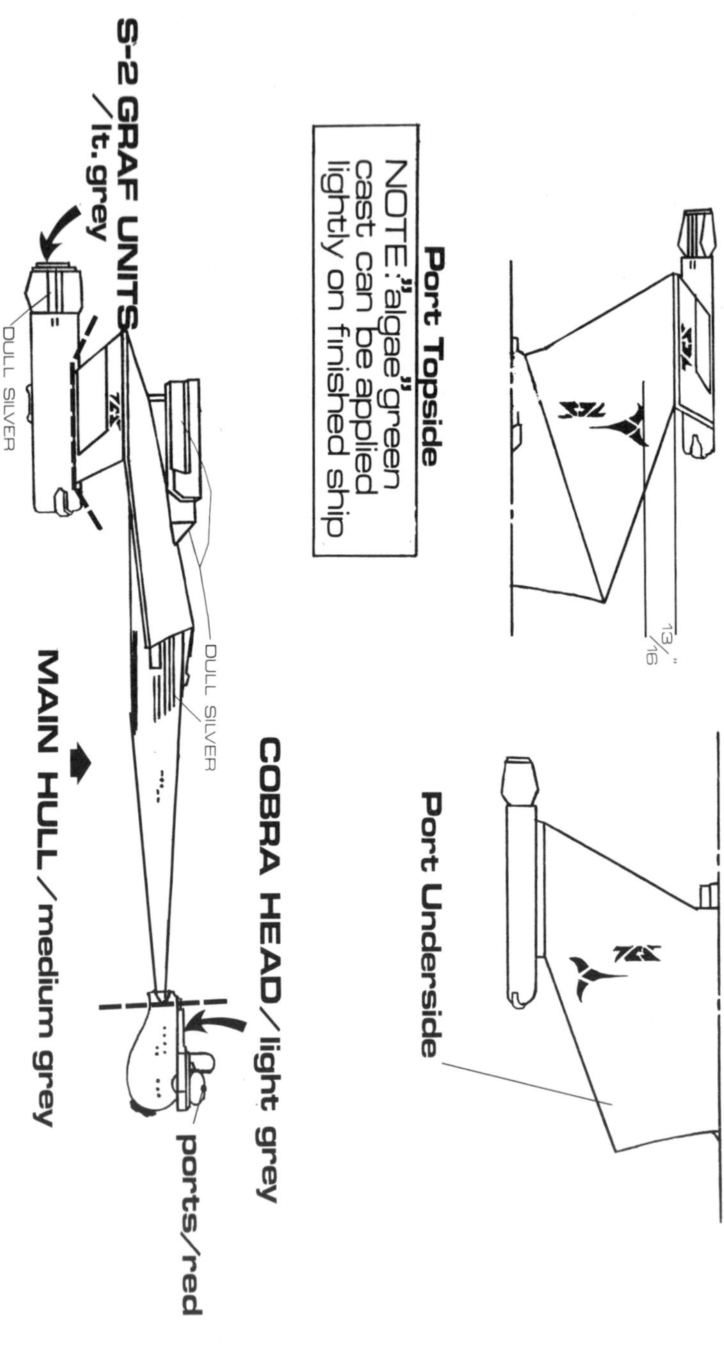

Romulan Bird-of-Prey

The Romulan Bird-of-Prey is another example of excellent spacecraft design. The bird design on the bottom of the ship was unique, and reflected a culture that chose to add symbols to their ships of war, just as the "Flying Tigers" of World War II.

The miniature of this ship was about a yard across and it's suggested that it was made of basswood, plastic, brass, and some plexiglas. This ship appeared in only two episodes, "Balance of Terror" and "The Deadly Years" of which the same stock special FX footage was used for both episodes. Unfortunately, close-ups of this beautiful ship were not done very well, so obtaining color photos, especially of the topside, are practically non-existent. The windows were lighted from a single light source as were the nacelle domes. The windows never appeared on film. The AMT version of this ship is not as accurate as the other Star Trek kits. Since the original miniature was lost or destroyed, AMT had little to go on, so they improvised areas that were "unknown". One example is the tips of the nacelles; AMT places a "ball" where only a half-sphere is necessary. Also, the underside became contoured at the rear, where the tail feathers are. The raised part (bridge) is not proportioned right, as well as some angle problems of the leading edge of the craft. However, considering what AMT had to work with, this kit, with some work, can be a fine model. This part of SFAM #2 will attempt to give some ideas on how to modify your Bird-of-Prey kit. Properly lighted and painted, this model can be quite impressive. For the perfectionists, to accurately model this craft would require to build from scratch.

1. THE ROMULAN BIRD-OF-PREY IS ONE OF THE EASIEST MODELS TO BUILD. IT TAKES A RELATIVELY SHORT TIME TO ASSEMBLE, DUE TO THE ABSENCE OF SMALL PARTS.

2. SHOWN AT RIGHT ARE THE TOTAL NUMBER OF PARTS: 16. CHECK FOR ANY MISSING PARTS. ALL PARTS SHOULD RELEASE QUITE EASILY FROM THE SPRUES. HOWEVER, A KNIFE MAY BE USEFUL IN RELEASING SOME PARTS. THERE SHOULD ALSO BE A DECAL SHEET AND A SET OF INSTRUCTIONS.

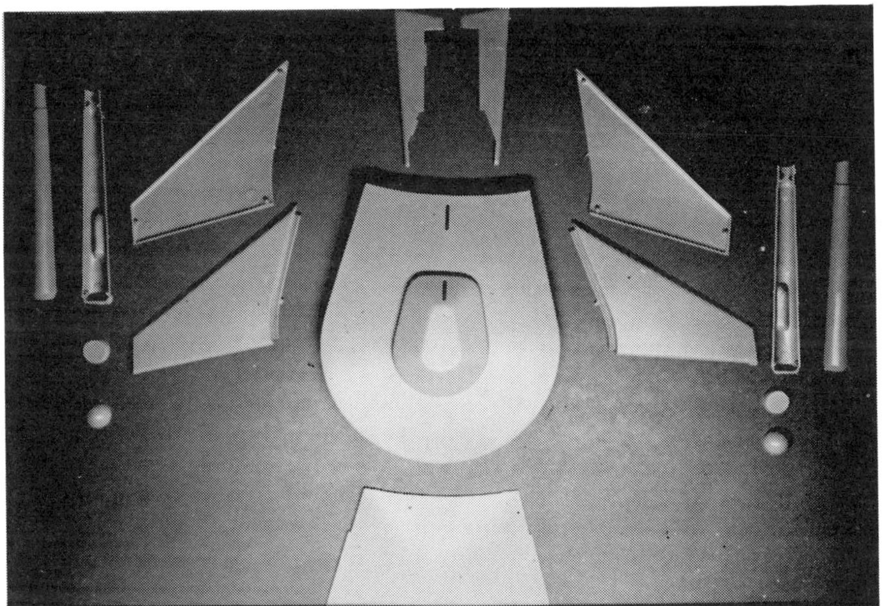

REFER TO PAGE AP:02:33.

3. DUE TO THE EASE OF ASSEMBLY OF THIS KIT IN ITS ORIGINAL FORM, THE MODIFICATIONS WILL BE EMPHASIZED HERE. FIRST OF ALL, IF YOU CHOOSE TO LIGHT THE NACELLES, THE HALF-SPHERES WILL NOT BE REQUIRED AND CAN BE USED ON YOUR STARSHIP AS THE NEAR-CORRECT ENLARGED VERSION OF "THOSE BALLS AT THE REAR OF THE NACELLES", AS SHOWN (WITH SOME TRIMMING AND PUTTY).

4. FIRST OF ALL, ACQUIRE A PLEXIGLAS ROD OF 5/8" DIAMETER. TAKE THE HALVES OF THE NACELLES AND DRILL OUT THE FLAT AREAS AS SHOWN. DRILL OUT AN AREA THAT WILL FIT SNUGLY AROUND THE ROD WHEN THE HALVES ARE FITTED TOGETHER. THIS WILL TAKE SOME TIME AND PATIENCE, SO BE CAREFUL.

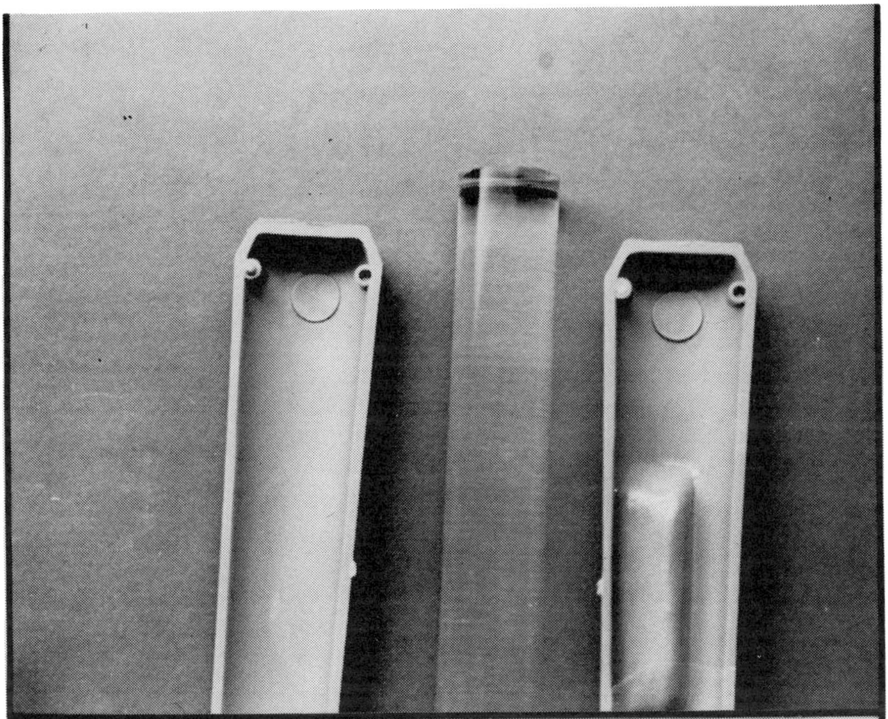

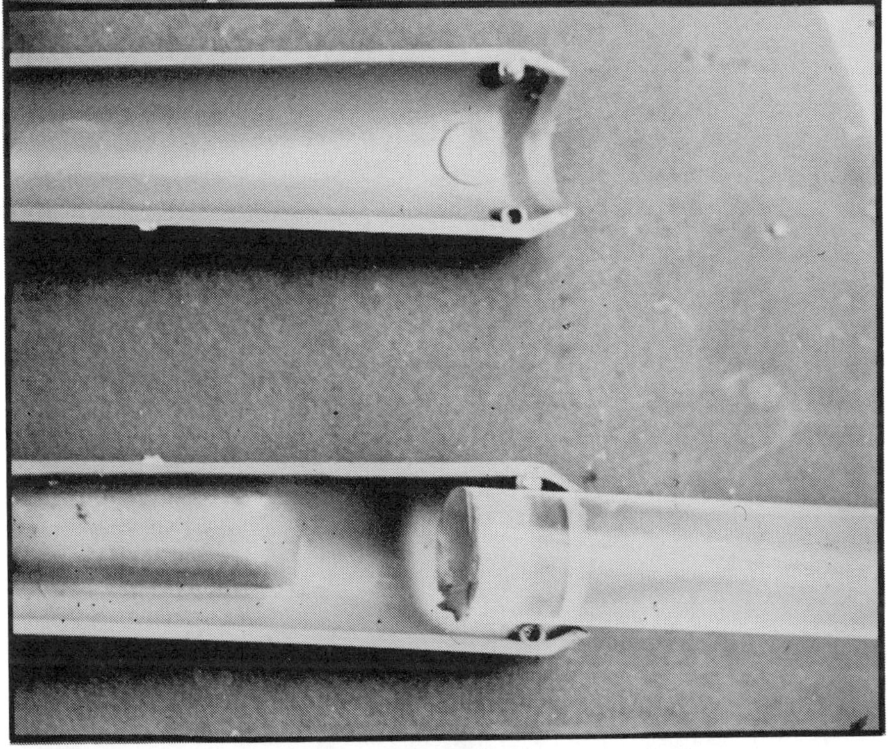

5. AFTER THE NACELLES HAVE BEEN SUCCESSFULLY DRILLED OUT, PLACE THE ROD INTO THE NACELLE END AND MEASURE THE DISTANCE DESIRED TO PROTRUDE OUT OF THE END, TO BE ROUNDED. THE PIECE SHOWN IS 3/4" LONG. THE ROUNDED PART SHOULD BE ABOUT 1/4", AS SHOWN. FIRST, WITH A FILE AND SANDPAPER, ROUND OFF THE ROD TIP, THEN SAW OFF 3/4" FROM THE TIP.

6. ONCE THE PIECE OF PLEXIGLAS ROD HAS BEEN CUT WITH A HACK OR RAZOR SAW, IT CAN BE DRILLED OUT WITH A HAND DRILL AND 3/16" OR 1/4" BIT. REMEMBER TO HOLD THE PLEXIGLAS IN PLACE WITH A LARGE VISE WHEN SAWING AND/OR DRILLING. THE HOLE SHOULD BE DRILLED ABOUT HALFWAY INTO THE PIECE, AS SHOWN.

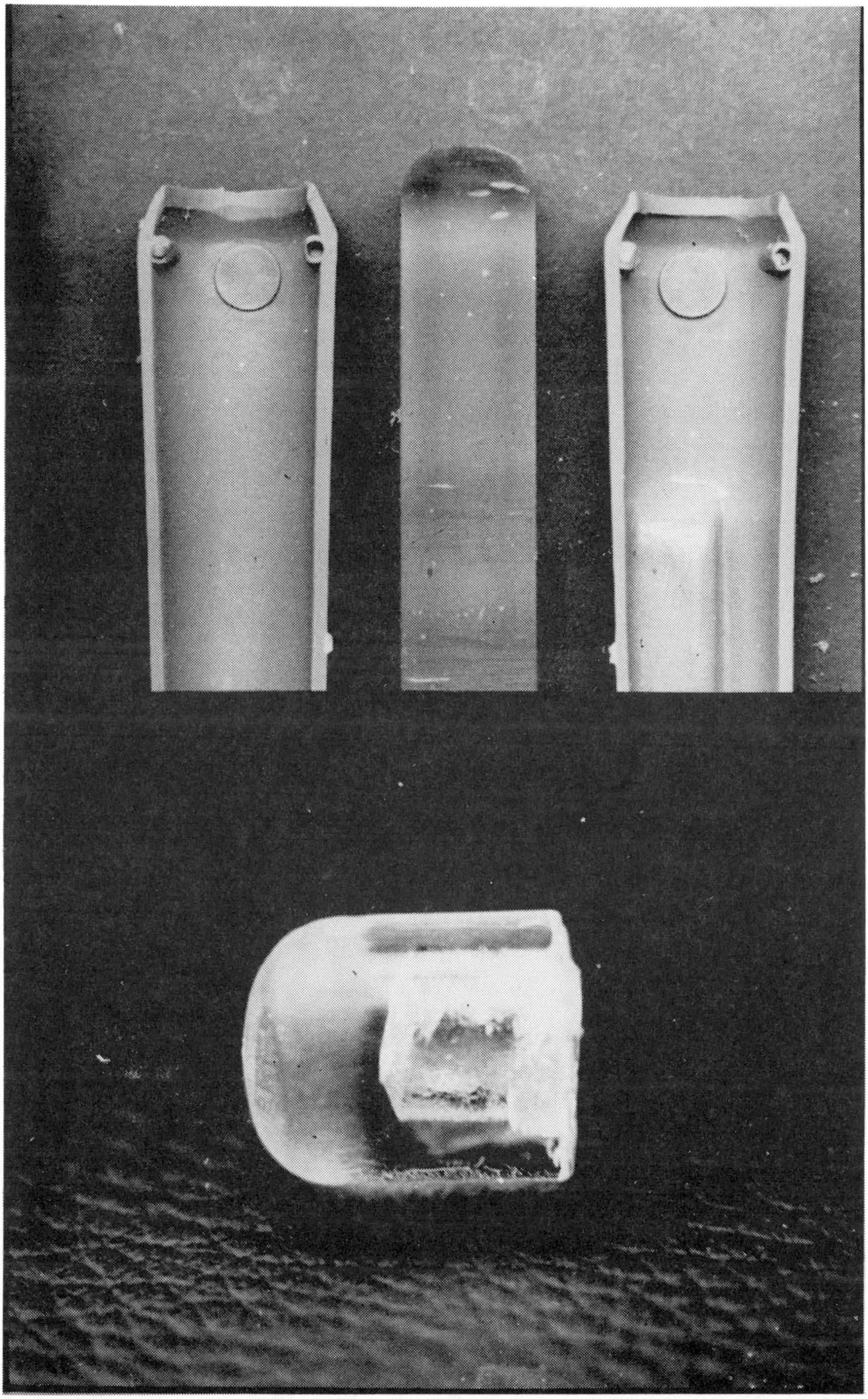

7. PLACE THE FINISHED PIECE INTO ONE HALF OF THE NACELLE HALVES. THIS HALF SHOULD BE GLUED TO A TOP HALF OF A SUPPORT PYLON. THE JOINED PIECES SHOULD BE DRILLED TO PROVIDE A HOLE FOR THE WIRES, AS SHOWN. CONNECT THE LAMP (RADIO SHACK) TO THE LEADING WIRES (ABOUT 1 FT) THAT WILL RUN THROUGH THE SUPPORT PYLON (WING) AND INTO THE SHIP'S HULL. THEN ATTACH THE OTHER HALF OF THE NACELLE, AND THE OTHER HALF FOR THE SUPPORT PYLON. REPEAT FOR THE OTHER NACELLE. SAND OR FILE THE SEAMS OF THE NACELLES AND PYLONS AND APPLY BODY PUTTY WHERE NECESSARY.

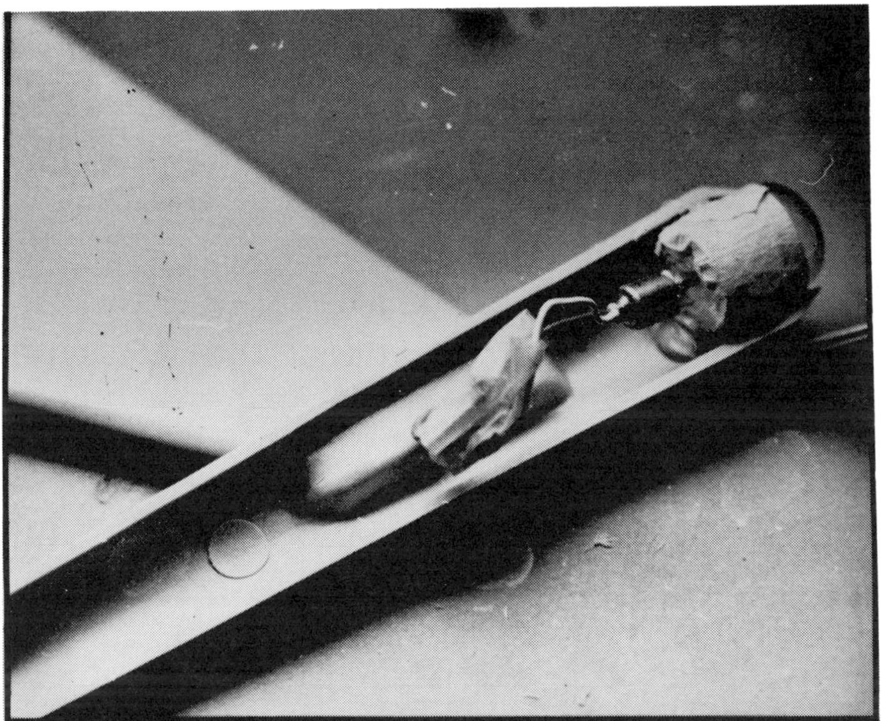

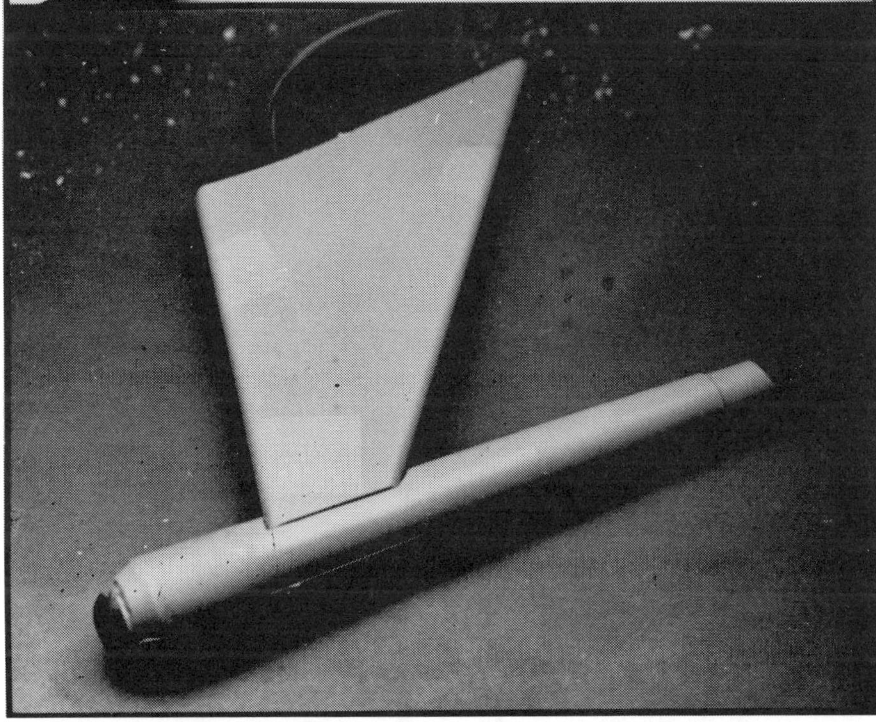

8. THE TOPSIDE AND UNDERSIDE ARE SHOWN AT RIGHT. THE DECAL IS ESPECIALLY FRAGILE AND TEARS EASILY. TO MAKE IT EASIER, COVER THE UNDERSIDE WITH WATER, THEN SOAK THE BODY OF THE BIRD DECALS. LAY THE DECALS ON TOP OF THE MODEL AND SLIDE THE DECAL OFF ONTO THE SHIP. BY WORKING TOWARDS THE ENDS, APPLY THE DECALS, ONE AT A TIME, SLIGHTLY OVERLAPPING. LASTLY, WHEN DRY, COAT WITH MATTE SPRAY.

9. THE MAIN HULL CAN BE LIGHTED AS WELL. DRILL OUT THE PLACES INDICATED BY THE DARK DOTS. EITHER USE FIBER OPTIC AS EXPLAINED IN AP-KBC #8, OR A SINGLE LIGHT SOURCE, WITH SCOTCH TAPE APPLIED TO THE INSIDE TO DIFFUSE THE LIGHT. THE VOLTAGE OF THE NACELLE LIGHTS CAN BE 1.5, 6 OR 12 VOLTS. IF A SINGLE LIGHT SOURCE IS DESIRED, TRY TO CHOOSE ONE THAT IS RATED THE SAME VOLTAGE AS THE NACELLE LAMPS. IF NOT, YOU MAY BURN OUT THE LAMPS. BATTERIES OR A VOLTAGE ELIMINATOR CAN BE USED TO SUPPLY POWER.

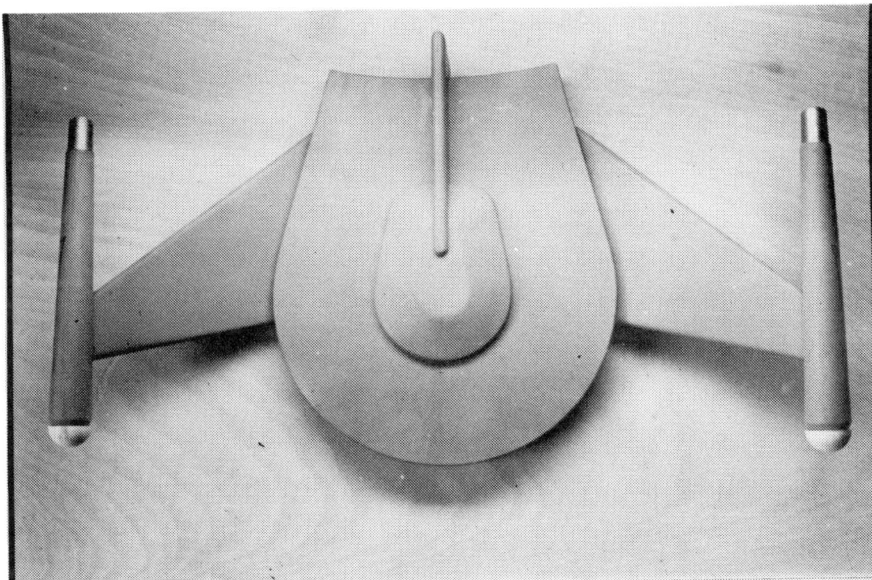

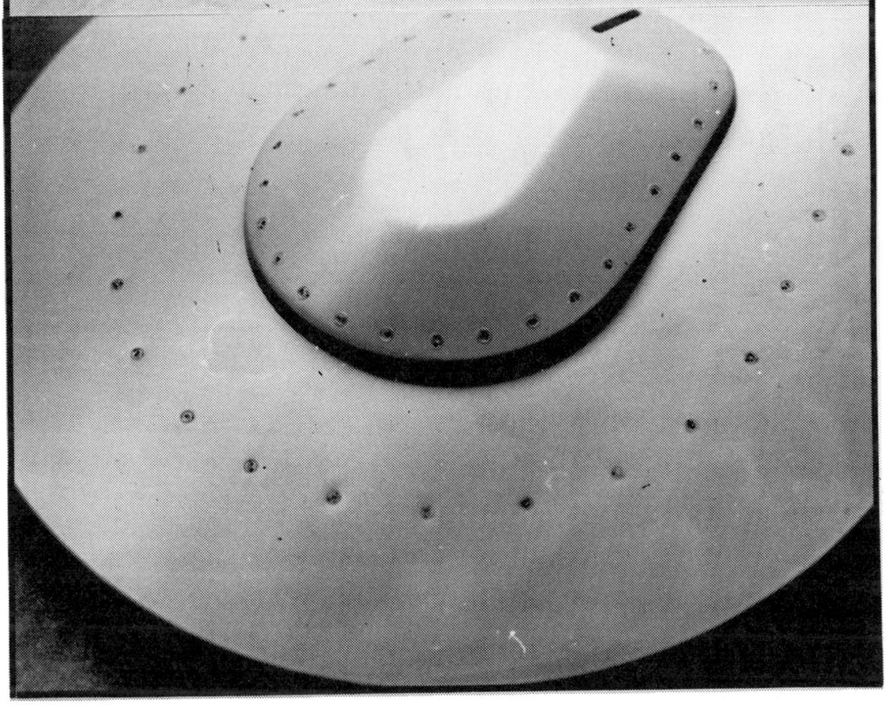

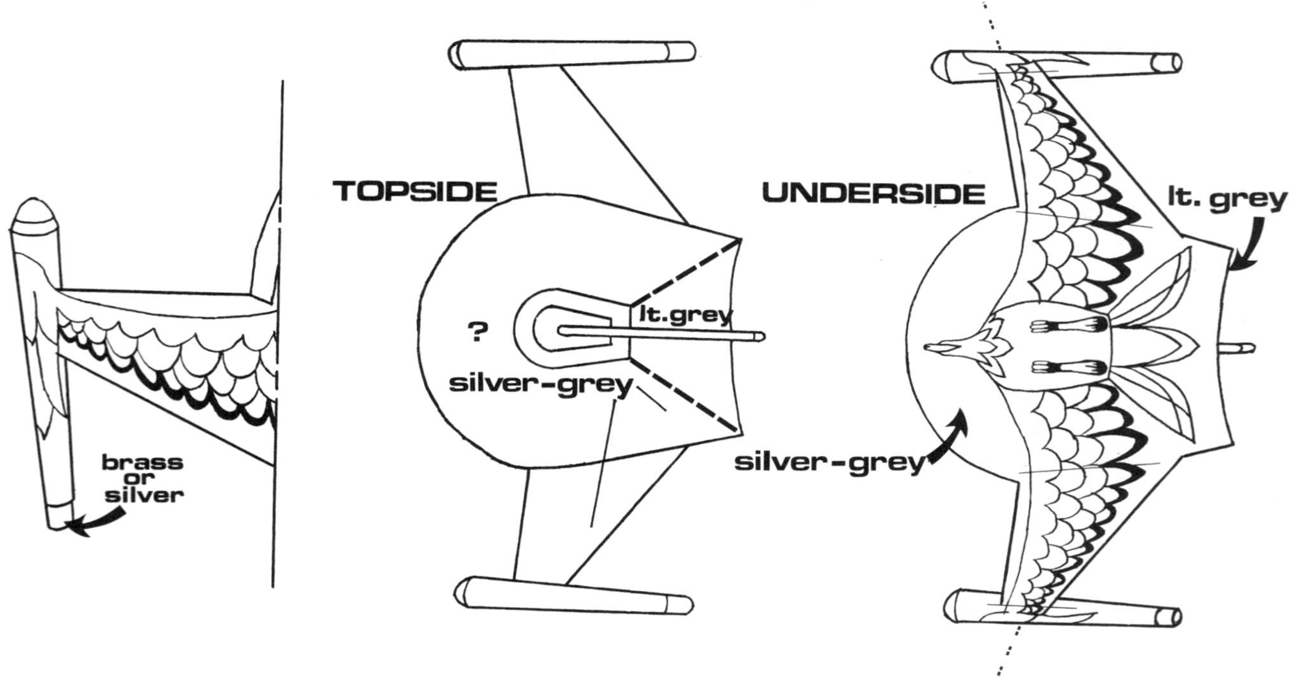

PAINTING SCHEMATIC FOR ROMULAN BIRD·OF·PREY RBP-2

DISPLAY BASE SECTION

AFTER YOUR SHIP HAS BEEN COMPLETED, TO EFFECTIVELY DISPLAY IT YOU MAY DECIDE ON ADDING A BASE. SOME OF YOU MAY DECIDE TO HANG IT ON A STRING FROM THE CEILING, BUT IF YOU'RE USING THE LIGHTING TECHNIQUES, A BASE MAY BE REQUIRED TO HOUSE THE POWER SUPPLY. TO BUILD A SIMPLE BASE FOLLOW THE PROCEDURE AS FOLLOWS

1. ACQUIRE A BRASS ROD MEASURING 3/8" TO 5/8" DIAMETER AND ABOUT 10" LONG.

2. ALSO ACQUIRE A ROUND WOODEN BASE, PREFERABLY OAK, TO INSTALL THE ROD INTO. THESE BASES ARE AVAILABLE AT MANY HOBBY SHOPS OR LUMBER STORES AND ARE KNOWN AS "DECOUPAGE PLAQUES".

3. DRILL A HOLE SLIGHTLY SMALLER IN DIAMETER TO THAT OF THE SUPPORT ROD, INTO THE BASE. CHOOSE A PLACE 1-1/2" OUTWARDS FROM THE CENTER TO CREATE A COUNTERBALANCE. WITH A LARGE VISE GENTLY PUSH THE ROD INTO THE BASE BY TIGHTENING THE VISE. BE CAREFUL NOT TO BEND THE ROD.

4. THE BASE CAN BE STAINED WITH A LAQUER STAIN OR BLACK OR BROWN SHOE POLISH IF SO DESIRED.

5. IF YOU ARE USING ELECTRONICS OR LIGHTING, THE WIRES CAN BE RUN DOWN THE BRASS ROD. DRILL A HOLE, ALSO SLIGHTLY SMALLER INTO THE SHIP TO PROVIDE A SNUG FIT. CONNECT THE WIRES AT THE BASE TO A 9 VOLT BATTERY ELIMINATOR.

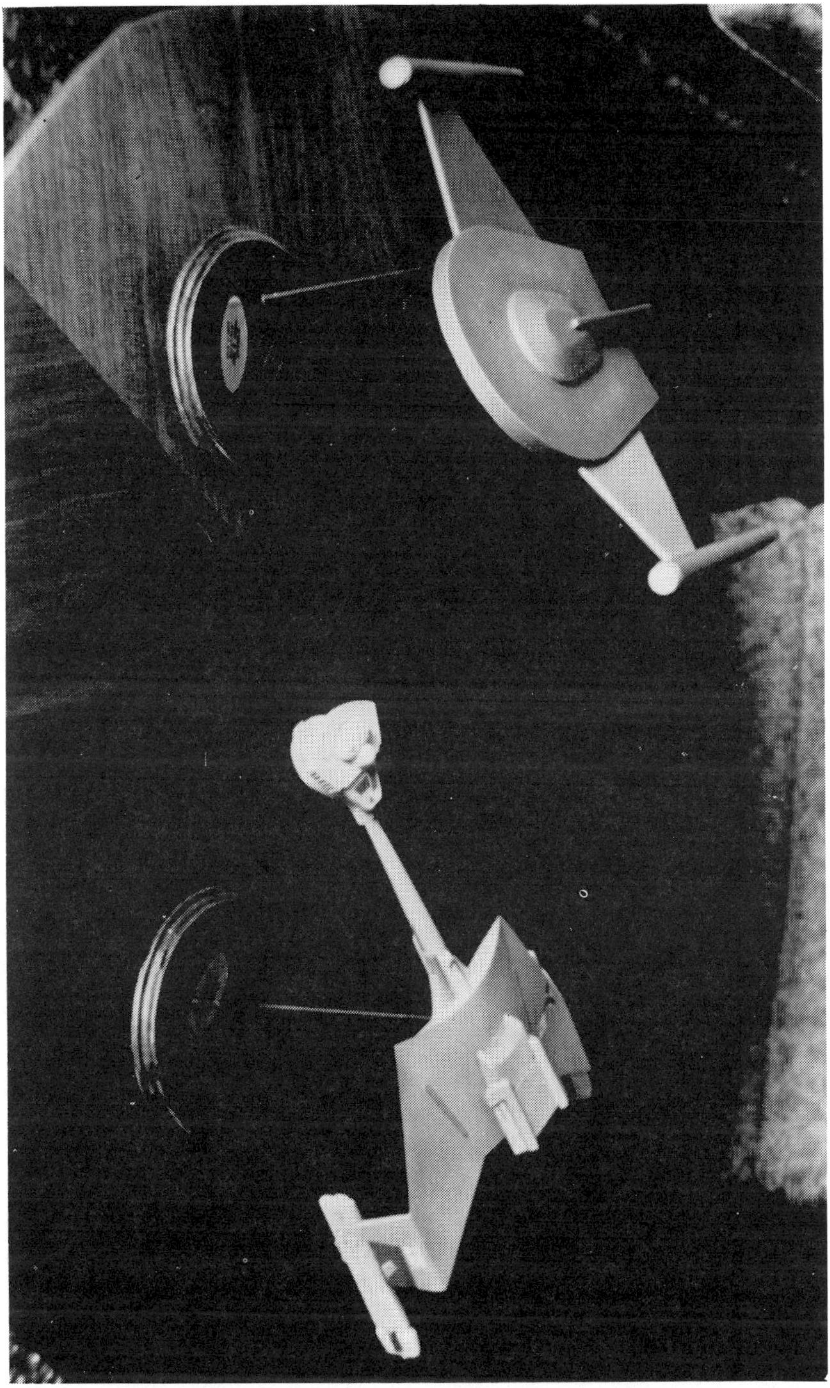

TO BATTLE DAMAGE YOUR SHIP THE FOLLOWING PAGES ILLUSTRATE ONE SUCH ATTEMPT. BELOW IS THE PROCEDURE TO FOLLOW TO ACHIEVE THIS PARTICULAR EFFECT.

1. PICTURED BELOW IS A REVERSE SHOT OF AN UNSCATHED S-2 GRAF UNIT. TO START, CUT WHERE THE DOTTED LINES SHOW WITH A DREMEL TOOL AND SAW BLADE.

2. REMOVE THE FRONT SECTION AND INSERT JUNK PIECES OF OTHER MODELS TO SIMULATE "GUTS". SHOWN IS A PIECE FROM AN N SCALE MODEL RR SIGNAL TOWER. THINGS WITH RIVETS OR EXTREME DETAIL WORK BEST.

3. ALSO ADD PIECES OF FOIL FROM CUPCAKES. THIS FOIL IS EXTREMELY THIN AND CRINKLES WELL. FIRST COMPRESS A BALL OF FOIL TIGHTLY, THEN TEAR IT IN HALF. USE PIECES OF THIS, GLUED IN STRATEGIC PLACES TO SIMULATE SHRAPNEL.

4. TO ADD TO THE EFFECT, USE A HOT SOLDERING IRON TO PUT NICKS FROM PHASER BLASTS INTO THE HULL'S SURFACE. USING THE DREMEL TOOL BY NICKING IT ON THE HULL ALSO WORKS WELL (SEE ARROWS).

5. AFTER YOU HAVE COMPLETED ADDING THE DETAILS, SPRAY THE ENTIRE DAMAGED AREA FLAT BLACK OR GRAPHITE. THEN SPRAY, AT A 45^{0} ANGLE, THE ORIGINAL SHIP COLOR ON TOP OF THE BLACK. THIS GIVES A RECESSED LOOK AND ACENTUATES THE DAMAGE, AS SHOWN. PAINT SURFACE DETAILS SILVER. FINALLY, SPRAY WITH A PROTECTIVE COAT OF MATTE SPRAY.

IN THIS BOOKLET:

- BASIC ASSEMBLY OF AMT KITS
- ADDING A BASE
- INSTALLING WORKING LIGHTS
- SPECIAL PAINTING TECHNIQUES
- ADDING CUSTOM BATTLE-DAMAGE
- PLUS MUCH MORE!!

2

StarFleet Assembly Manual 3

ADVANCED ASSEMBLY PROCEDURE FOR HEAVY CRUISER STARSHIPS

INCLUDES INSTRUCTIONS FOR

DIGITAL ELECTRONICS SYSTEM

LS/100A

ADVANCED ASSEMBLY PROCEDURE (A)APHC-3

CONCEIVED, WRITTEN and PHOTOGRAPHED by:

Paul M. Newitt

ELECTRONIC TECHNICAL ASSISTANCE and LS//100A ASSEMBLY SECTION WRITTEN by:

Terry Whittier

SENIOR ELECTRONIC LAB TECHNICIAN / IBM — San Jose

ARTWORK: NACELLE REPAIR IN SPACE rendered by:

Al Sirois

SUPERDETAILING/PARTS DIAGRAMS by:

David Merriman

D & E MODELS

SPECIAL MATERIAL PROVIDED by:

Franz Joseph

UNITED FEDERATION REPRESENTATIVE

SPECIAL THANKS TO THESE PEOPLE FOR THEIR CONTRIBUTIONS:

MOM & DAD, MARK D. NEWITT, CLARENCE NICODEMUS, A.J. DESMARETZ RICHARD G. VAN TREUREN, JOHN MOORE, RICK NEIDIG, JOHN F. DONIGAN JAMIE MURPHAY, FRANK CLAUNCH, DAN MOORE, EUGENE DORR, JAMES FULKERSON, JIM ROBERTSON, DAN TIRTAWINATA, FRANK KURSON, WALTER IRWIN, CHUCK WEISS, & GEOFFREY MANDEL.

LS//100A system conceived by Paul M. Newitt
engineered by Terry Whittier

StarFleet Division

LS/100A Digital Electronics System

tSD 7808.20

Dear Modelers/Fans:

StarFleet Assembly Manual Three is the third and final of the SFAM series, outlining the most comprehensive modeling techniques on building the AMT U.S.S. Enterprise ever presented. With this booklet, any variation of modifications can be made on your Heavy Cruiser Starship, including lighted ports, sensor arrays, working running lights, homing beacon, and warp engines.

Over a year of research and development, together with suggestions, comments and hints from fans, has gone into this booklet and electronics system. Terry Whittier and I worked diligently on the LS/100A Digital Electronics System to provide you with the most advanced, yet easy-to-install lighting system for your Heavy Cruiser Starship. Timing rates for the running lights, homing beacon and warp engines were carefully calibrated and required research from viewing parts of the episodes that few check in such detail. Four different arrangements for the fiber optic were also worked out to develop a single, simplified method of lighting the ports most effectively. Other details which are too numerous to mention, were researched as well.

The modeling techniques used in this manual are basically my own with some suggestions and alternative methods listed. Each modeler has his/her own technique, so I strongly suggest that you read the entire booklet in detail to determine your own method of how to build this ship, and to modify any steps. This advanced assembly procedure was designed for those with some experience with this kit, as well as in general model building and electronics (recommended age - 12 years to adult). It's also important to determine how much to spend on this project. The ship pictured in its completed form was about $40.00 not including tools. The AMT kit has many possibilities to becoming a beautiful miniature, in fact, it was used in episodes like: "Doomsday Machine" and "Trouble With Tribbles". Your ship, when completed, will certainly rival those studio miniatures!

Spacecraft modeling is finally being realized as a legitimate part of scale modeling, as well as a strong part of ST/SF fandom. It is the hope of StarFleet Division to promote these models seriously as scale miniatures rather than just "toys", and to continue development of such innovative designs. Equally as important is StarFleet Division's devotion to each and every fan, and to continue the personal response to everyone who writes.

Last, but not least, I'd like to thank Franz Joseph, United Federation Representative, for his thoughtful assistance with providing the Marking and Painting Specification - Class I Starships. His contribution is truly an integral part of this manual.

If you have any unsolvable problems with this advanced procedure, feel free to write me. Thank you for your patience and support.

Peace and Long Life

Paul M. Newitt

/pmn

LS//100A Digital Electronics System™

ASSEMBLY

CONSTRUCTION OF THE LS//100A CIRCUIT WILL CONSIST OF THE FOLLOWING STEPS:

1. ACQUIRING AND CHECKING ALL PARTS AND TOOLS TO BE SURE ALL MATERIALS ARE AT HAND TO BEGIN BUILDING.
2. ASSEMBLING AND SOLDERING THE BOARD CIRCUITRY.
3. PREPARING AND ATTACHING ALL WIRES TO THE BOARD.
4. ADDING THE L.E.D.s (LIGHT EMITTING DIODES).
5. POWERING UP THE BOARD AND CHECKING THAT IT IS FULLY FUNCTIONAL BEFORE INSTALLING IT IN THE SHIP.

PLEASE READ AHEAD AT THIS POINT AND FAMILIARIZE YOURSELF WITH THE MATERIALS AND ASSEMBLY PROCEDURES DISCUSSED ON THE NEXT FEW PAGES BEFORE BEGINNING ACTUAL CONSTRUCTION.

STEP 1A - TOOL LIST (ELECTRONIC CONSTRUCTION)

* * - SMALL "THIN NOSE" PLIERS (ALSO ACCEPTABLE: "CHAIN NOSE", "NEEDLE-NOSE" PLIERS)
* - SOLDERING IRON WITH PENCIL TIP (25-30 WATT IRON -- RADIO SHACK #64-2070 ACCEPTABLE)
* - WIRE STRIPPERS/CUTTERS
* - SAFETY GOGGLES OR FACE SHIELD FOR EYE PROTECTION DURING SOLDERING
* * - SMALL DIAGONAL CUTTERS (FINGER NAIL CLIPPER ACCEPTABLE)
* - SUPER GLUE

STEP 1B - PARTS LIST (ELECTRONIC)

- INTEGRATED CIRCUITS, DIP (DUAL IN-LINE) STYLE
  - *1 - 556 DUAL TIMER
  - *1 - 7490 DECADE COUNTER
  - *1 - 7441 BCD TO DECIMAL DECODER/DRIVER
- CAPACITORS - ELECTROLYTIC, AXIAL LEAD TYPE
  - *1 - 10 MICRO-FARAD (C1)
  - *1 - 1 MICRO-FARAD (C2)
- RESISTORS - 1/4 WATT, 5%
  - *2 - 12K OHM (R9, R8)
  - *1 - 100K OHM (R6)
  - *1 - 150K OHM (R7)
  - *5 - 300 OHM (R1 THRU R5)
  - OPTIONAL: 1 - 50K OHM PRINTED CIRCUIT POTENTIOMETER (REPLACES R8)
    1 - 500K OHM P.C. POT (REPLACES R7)
- L.E.D.s
  - *10 - JUMBO CLEAR (NACELLE LIGHTS)
  - * 3 - MINIATURE YELLOW (A3 - A5)
  - * 1 - MINI RED (A1)
  - * 1 - MINI GREEN (A2)
- * - 6 FT. - ROSIN CORE SOLDER, 22 OR 20 GAUGE
- 4 - 6 VOLT, 25 MILLIAMP INCANDESCENT LAMPS WITH LEADS (R.S. #272-1140) (L1 - L4)
- * - 1-1/2 FT. - 12 CONDUCTOR FLAT COMPUTER CABLE
- * - 10 FT. - 26 GAUGE STRANDED, TWISTED PAIR WIRE
- 1 - 110V A.C. TO 6V D.C. POWER SUPPLY, RATED 250 MILLIAMP CURRENT (K-MART MULTI-VOLTAGE AC ADAPTOR OR SEARS #61K-2147) (CAUTION - WHEN USING A MULTI-VOLTAGE POWER SUPPLY, MAKE SURE IT IS SET TO 6V!)
- 1 - 1/4" PHONE PLUG (R.S. #274-325 ADAPTER-MINIATURE PHONE TO 1/4" PHONE PLUG)
- 1 - 1/4" PHONE JACK (R.S. #274-252- 2-CONDUCTOR OPEN CIRCUIT PHONE JACK)
- * - 1 - DIODE, IN4001 OR IN4002

NOTE: ITEMS STARRED (*) ARE INCLUDED IN THE LS//100A DIGITAL ELECTRONICS KIT ALONG WITH THE PRINTED CIRCUIT BOARD AND NACELLE DISCS. PLEASE SEE BACK OF BOOK FOR MORE INFORMATION.

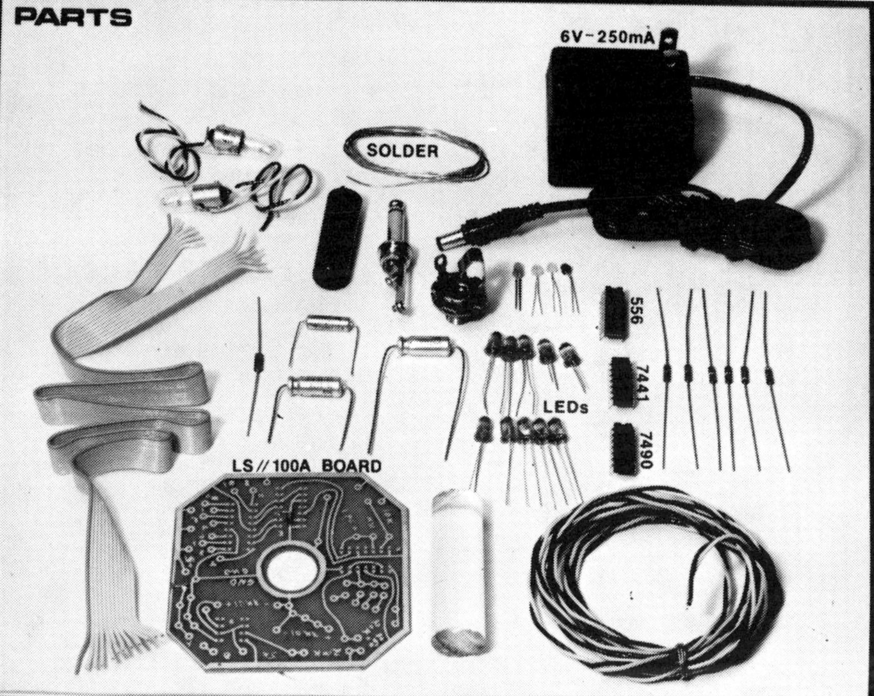
PARTS
6V~250mA
SOLDER
556
7441
7490
LEDs
LS//100A BOARD

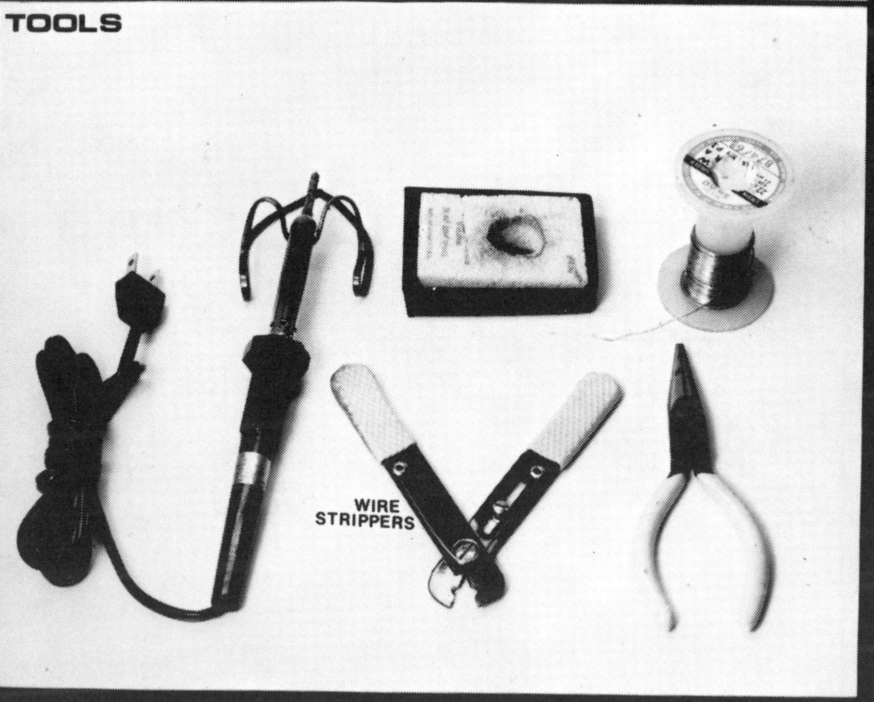
TOOLS
WIRE
STRIPPERS

LS//100A Digital Electronics System™ ASSEMBLY

THEORY OF OPERATION

THE FUNCTION OF THE LS//100A CIRCUIT IS DIVIDED INTO TWO PARTS: THE BLINKING RUNNING LIGHTS AND THE WARP DRIVE SIMULATOR.

THE 556 IS A DUAL TIMER, EACH HALF FULLY INDEPENDENT IN ITS OPERATION. ONE HALF OF THE 556 DIRECTLY DRIVES THE FIVE RUNNING LIGHTS, ACCORDING TO AN ON-OFF PULSE SET BY THE COMBINATION OF RESISTORS R6, R7, AND CAPACITOR C1.

THE OTHER HALF OF THE 556 GENERATES PULSES (SET BY R8, R9, and C2) THAT FEED INTO THE 7490 DECADE COUNTER. THIS COUNTER CHIP CONTINUOUSLY COUNTS THE INCOMING PULSES, STARTING WITH 0 AND ENDING WITH 9, THEN STARTING OVER AGAIN. THE COUNT AT ANY INSTANT APPEARS ON THE OUTPUT PINS OF THE 7490 IN BCD CODE (BINARY CODED DECIMAL CODE). THE BCD CODE FEEDS INTO A 7441 BCD TO DECIMAL DECODER/DRIVER. THIS 7441 INTERPRETS THE CODE AND GROUNDS THE APPROPRIATE OUTPUT PIN ON ITSELF. THE 10 OUTPUT PINS (CORRESPONDING TO THE 7490's 0 THROUGH 9 COUNT) OF THE 7441 ARE CONNECTED BY THE FLAT COMPUTER CABLE TO 10 LEDs IN THE NACELLES. THE WARP DRIVE EFFECT IS THEN CREATED BY LIGHTING UP 5 LEDs IN EACH NACELLE IN CONSECUTIVE SEQUENCE AROUND IN A CIRCLE. ONE NACELLE USES EVEN NUMBERS AND THE OTHER USES ODD NUMBERS, BUT THE EFFECT APPEARS SIMULTANEOUS AND CONTINUOUS. THE POSITIVE LEAD OF EACH OF THESE NACELLE LEDs IS TIED TO A POSITIVE 5V, AND WHEN THE OTHER LEAD IS GROUNDED BY THE 7441, CURRENT FLOWS THROUGH IT AND THE LED LIGHTS UP. (NOTE: THE CURRENT THROUGH DIODES MUST BE LIMITED - THIS IS DONE WITH A 300 OHM RESISTOR FOR THE RUNNING LIGHTS, BY THE LS//100A CIRCUIT IN THE CASE OF D1, AND BY THE 7441's INTERNAL CIRCUITRY FOR THE NACELLE LEDs).

THE INCANDESCENT LAMPS (L1-L4) ARE CONNECTED DIRECTLY ACROSS THE INCOMING 6V AND PROVIDE A STEADY LIGHT TO ILLUMINATE THE PRIMARY HULL DOMES AND SHIP'S WINDOWS.

DIODE D1 BOTH PROTECTS THE CIRCUIT FROM BEING HOOKED UP TO POWER BACKWARDS (SINCE DIODES ONLY ALLOW CURRENT TO FLOW IN ONE DIRECTION) AND DROPS THE VOLTAGE FOR THE INTEGRATED CIRCUIT CHIPS AND LEDs DOWN CLOSER TO A STANDARD 5 VOLTS.

ADVANCED OPTIONS

TO VARY THE SPEEDS OF NACELLE ROTATION AND RUNNING LIGHT FLASHING, SUBSTITUTE A 50K OHM POTENTIOMETER FOR R8 AND A 500K OHM POT FOR R7. YOU CAN EASILY ATTACH THEM BY SOLDERING ONE WIRE OF A SECTION OF TWISTED PAIR WIRE TO THE CENTER PIN OF A POT AND THE OTHER WIRE TO EITHER OF THE END PINS. THEN SOLDER THE OTHER END OF THE PAIR OF WIRES WHERE THE ENDS OF THE REPLACED RESISTOR WOULD GO.

ALSO, YOU CAN BEEF UP THE INTERIOR LIGHTS BY RUNNING HIGHER AMPERAGE LIGHT BULBS OFF A SEPARATE POWER SUPPLY. THUS, YOU WOULD HAVE A 6V SUPPLY FOR THE DIGITAL CIRCUITRY, AND A 6, 9, OR 12 VOLT SUPPLY STRICTLY FOR THE BULBS. IF YOU DO THIS, USE A STEREO PHONE PLUG AND JACK INSTEAD OF MONO. RUN THE GROUNDS OF BOTH POWER SUPPLIES TOGETHER THROUGH ONE OF THE LINES, AND THE TWO POSITIVE LINES SEPARATELY THROUGH THE OTHER TWO. BE SURE THE GROUNDS DON'T TOUCH EITHER OF THE POSITIVE LINES, AND NEVER TIE THE TWO POSITIVE LINES TOGETHER, EVEN IF THEY ARE THE SAME VOLTAGE.

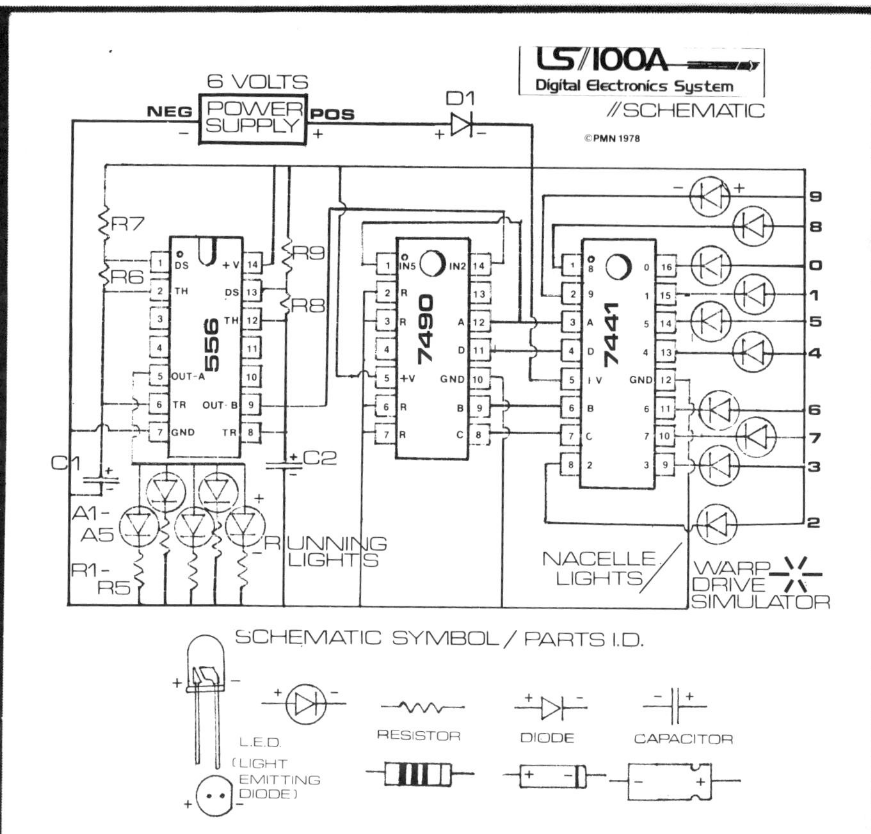
LS/100A
Digital Electronics System
//SCHEMATIC
©PMN 1978
6 VOLTS
NEG
POWER SUPPLY
POS
D1
R7
R6
R9
R8
556
7490
7441
C1
C2
A1-A5
R1-R5
RUNNING LIGHTS
NACELLE LIGHTS
WARP DRIVE SIMULATOR
SCHEMATIC SYMBOL / PARTS I.D.
L.E.D. (LIGHT EMITTING DIODE)
RESISTOR
DIODE
CAPACITOR

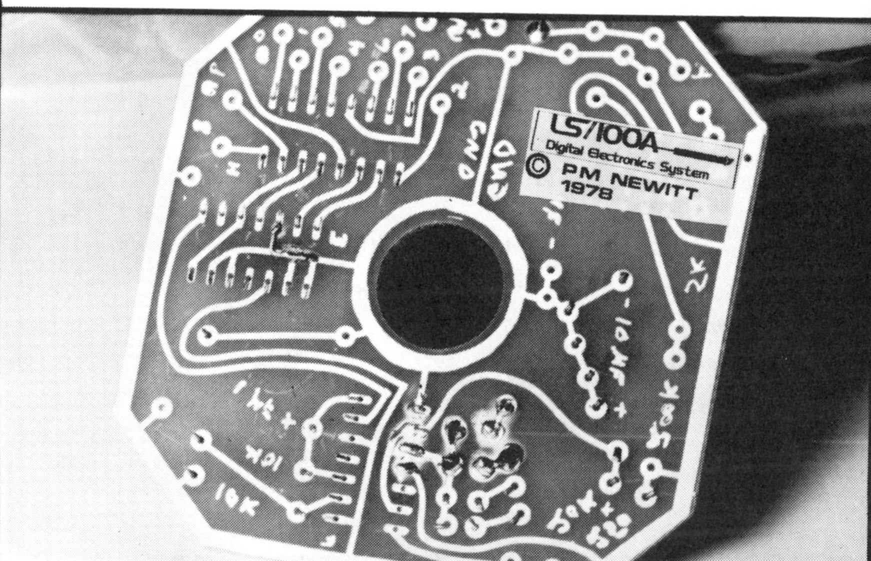
LS/100A
Digital Electronics System
© PM NEWITT 1978
GND

LS//100A Digital Electronics System™ ASSEMBLY

SOLDERING -- PLEASE READ CAREFULLY

CAUTIONS:

- SOLDERING IRONS REACH 700 TO 800 DEGREES. AVOID USING NEAR GLUE OR OTHER FLAMMABLE SUBSTANCES. NEVER LEAVE THE ROOM WHILE THE IRON IS STILL HOT OR STILL PLUGGED IN...WAIT 10 MINUTES AFTER UNPLUGGING. KEEP SOLDERING IRON AWAY FROM SMALL CHILDREN.
- WEAR SAFETY GOGGLES OR FACE MASK WHILE SOLDERING. HOT FLUX WILL SPIT, AND/OR WIRES CAN FLIP HOT, MOLTEN SOLDER INTO YOUR EYES.
- DO NOT HOLD IRON ON COMPONENTS, LED LEAD, I.C. PINS, WIRES, OR BOARD FOR MORE THAN 15 SECONDS AT A TIME. DAMAGE CAN RESULT FROM PROLONGED HEATING.

HOW TO SOLDER

TO SOLDER A COMPONENT LEAD OR I.C. PIN TO A PAD ON A PC (PRINTED CIRCUIT) BOARD, TOUCH THE TIP OF YOUR IRON TO BOTH THE LEAD AND PAD AT THE SAME TIME. LIGHTLY APPLY SOLDER TO THE JUNCTION OF IRON AND LEAD. A LITTLE SOLDER WILL MELT, RELEASING THE ROSIN FLUX IN ITS CORE. THIS FLUX WILL FLOW ALONG THE SURFACE OF METALS, CLEANING THE WAY FOR SOLDER. APPLY JUST ENOUGH SOLDER TO LIGHTLY COAT THE LEAD AND PAD, BUT NOT SO MUCH THAT IT FORMS A BLOB. AN OUTLINE OF THE LEAD SHOULD STILL BE VISIBLE BENEATH THE SOLDER.

"TINNING" MEANS TO PRE-COAT WITH SOLDER TO AID SOLDERING AT A LATER TIME. THE LS//100A PC BOARD IS PRE-TINNED TO MAKE SOLDERING MUCH EASIER, AND MANY LEADS ON ELECTRONIC PARTS ARE TINNED. INDIVIDUAL STRANDS OF STRANDED WIRE ARE OFTEN TINNED, AND YOU CAN TIN THE STRANDS TOGETHER AFTER STRIPPING THE END OF STRANDED WIRE TO BOND THEM TOGETHER FOR EASIER HANDLING. TO DO SO, STRIP ABOUT 1/4" OF INSULATION OFF THE END OF THE WIRE AND TWIST STRANDS FIRMLY TOGETHER. TOUCH THE WIRE WITH YOUR IRON NEAR THE INSULATION. THEN AS YOU STEADILY APPLY SOLDER, DRAW THE IRON DOWN THE EXPOSED WIRES AND OFF THE END. LATER, ALL YOU NEED TO DO TO JOIN THE ENDS OF TWO WIRES TOGETHER OR TO A PRE-TINNED LED LEAD, IS PLACE THEM SIDE-BY-SIDE AND HEAT WITH A CLEAN IRON UNTIL THE SOLDER FLOWS TOGETHER AND BETWEEN THE TWO.

CARING FOR YOUR SOLDERING IRON TIP: TIN BY APPLYING A COAT OF SOLDER TO THE END 1/4" TO 1/2" AS SOON AS IT WARMS UP ENOUGH TO MELT SOLDER THE FIRST TIME YOU PLUG IT IN. THE SECRET TO KEEPING A TIP IN GOOD CONDITION IS KEEPING IT CLEAN AND LIGHTLY TINNED WHILE IN USE AND AT REST.

REMOVE EXCESS SOLDER AND BURNT FLUX BUILD-UP BY POINTING THE IRON DOWN INTO A WASTEBASKET AND BRISKLY BRUSHING JUNK DOWN AND OFF THE TIP WITH A SMALL WAD OF PAPER.

DO NOT LEAVE THE IRON ON WHILE NOT BEING USED FOR A PROLONGED PERIOD. AFTER A WHILE, THE TIP WILL CORRODE AND LOSE ITS SHAPE BECAUSE THE COPPER OF THE TIP WILL SLOWLY DISSOLVE INTO ITS SOLDER COATING. WHEN THIS HAPPENS, YOU CAN RESHAPE BY LETTING THE IRON COOL COMPLETELY AND FILING IT DOWN TO BARE COPPER AND INTO THE SHAPE YOU WANT. THEN JUST RE-HEAT, TINNING AS SOON AS IT GETS HOT ENOUGH TO MELT SOLDER.

STEP 2 - LS//100A BOARD CONSTRUCTION

INSERT ALL CAPACITORS, RESISTORS, I.C.s AND THE DIODE IN BOARD, BEING CAREFUL TO OBSERVE PROPER POLARITIES AND ORIENTATIONS. SEE PHOTO. (NOTE: COMPONENTS AND WIRES ALL ARE INSERTED SO THEY REST ON THE BLANK SIDE OF THE BOARD AND THE LEADS STICK OUT THROUGH THE PRINTED CIRCUIT PADS.) BEND OVER ALL LEADS AND CUT OFF WITH NAIL CLIPPERS OR SMALL DIAGONAL CUTTERS SO LEADS DO NOT GO PAST EDGE OF PADS, BUT LEAVING ENOUGH LEAD TO SOLDER ONTO: ABOUT 1/16" TO 1/8" LONG. SEE DIAGRAM.

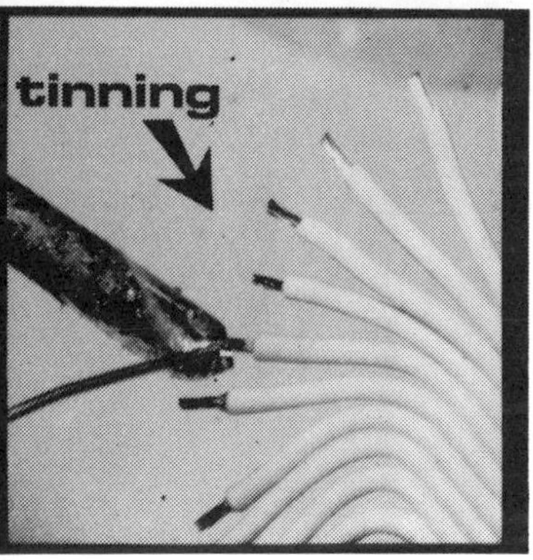

SOLDERING DIAGRAM

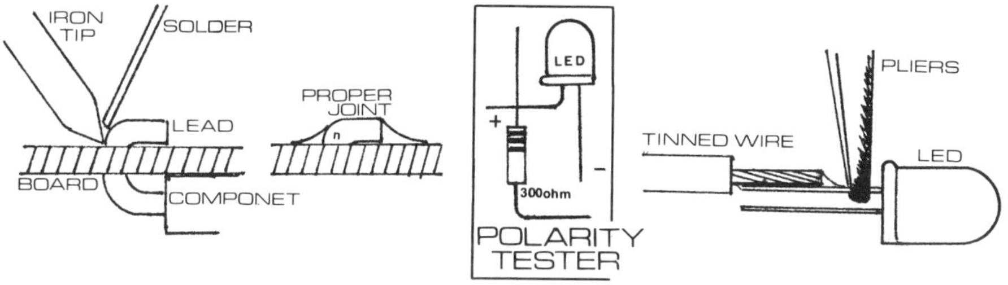

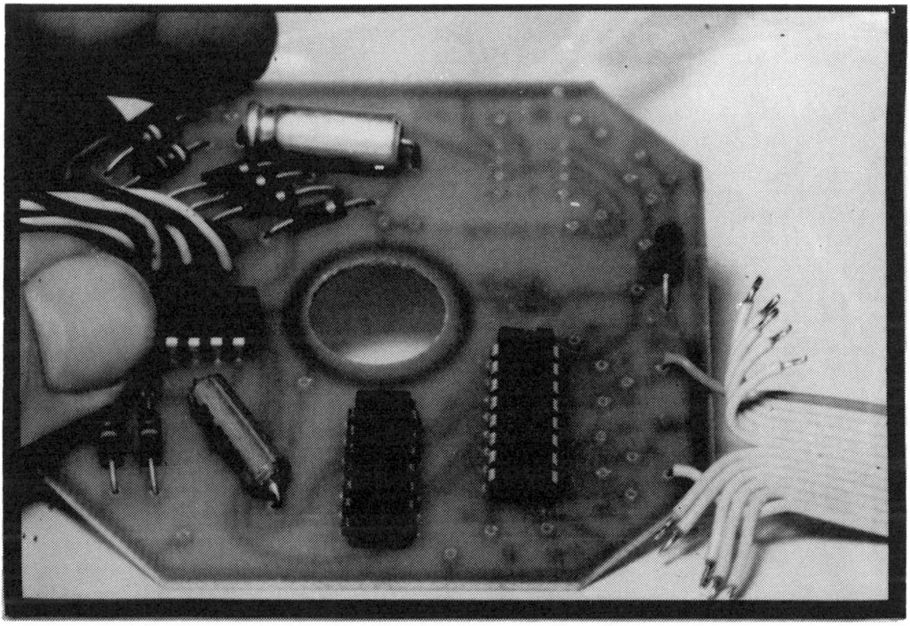

LS//100A Digital Electronics System TM

STEP 3 - WIRE PREPARATION AND SOLDERING

CUT ALL WIRES TO LENGTHS LISTED BELOW:
POWER INPUT LINE - 8" T.P. (TWISTED PAIR)
PRIMARY HULL RUNNING LIGHTS - 4 LENGTHS, 9" EACH T.P.
SECONDARY HULL HOMING BEACON - 12-1/2" T.P.
PRIMARY HULL LAMPS - NONE (EACH COMES WITH 4" LEADS)
SECONDARY HULL LAMPS - ONE 10" T.P.
FLAT COMPUTER CABLE FOR NACELLES - 18"

STRIP ABOUT 1/4" OFF ENDS OF ALL WIRES WITH WIRE STRIPPERS. TWIST STRANDS TOGETHER ON EACH AND PRE-TIN. SOLDER WIRES TO BOARD ACCORDING TO PHOTO.

NACELLE CABLE - FOR THE LAST 1-1/2" OF BOTH ENDS OF THE CABLE, SEPARATE INDIVIDUAL WIRES BY PULLING APART. AFTER TINNING, SOLDER ONE END OF CABLE TO BOARD ACCORDING TO NACELLE CABLE WIRING DIAGRAM BY PUSHING TINNED END OF EACH WIRE THROUGH FROM COMPONENT SIDE OF BOARD, BENDING OVER AND SOLDERING (SEE SOLDERING DIAGRAM).

RUNNING LIGHTS AND HOMING BEACON - CUT SMALL LED LEADS TO 1/2". SOLDER APPROPRIATE WIRES TO LEADS ONE AT A TIME BY GRASPING THE LEAD WHERE IT ENTERS THE LED WITH THE THIN NOSE PLIERS (WRAP A RUBBER BAND AROUND THE HANDLES OF THE PLIERS SO THEY WILL CLAMP ONTO THE LEAD FIRMLY BY THEMSELVES). BEND THE LEAD IN AN ARC SO IT FORMS A RIGHT ANGLE AND SOLDER THE PRE-TINNED WIRE ONTO IT. IF THERE IS TROUBLE GETTING A GOOD, UNIFORM BOND, ADD A LITTLE FRESH SOLDER TO THE LEAD AND TRY AGAIN WITH A CLEAN IRON. RUNNING LIGHT WIRES ALL GO TO PADS LABELLED A1 THROUGH A5 - BLACK WIRES GOING TO THE A SIDE PADS, AND THE LIGHTER WIRES GOING TO THE PADS ON THE NUMBER SIDE THAT ARE ALL CONNECTED TO EACH OTHER. THE BLACK LEAD ATTACHES TO THE NEGATIVE LEAD OF THE LED (THE LEAD ON THE SIDE OF THE LED WITH THE FLATTENED RIM - SEE SCHEMATIC SYMBOLS/PARTS IDENTIFICATION DIAGRAM).

PRIMARY HULL FILAMENT LAMPS - SOLDER TO PADS LABELLED L1 AND L2.

SECONDARY HULL LAMPS - SOLDER ONE END OF THE 10" LENGTH OF TWISTED PAIR WIRE TO PADS LABELLED L3. SOLDER THE OTHER END TO THE TWO REMAINING LAMPS BY SOLDERING THE BLACK WIRE TO THE RED WIRES OF THE LAMPS AND THE OTHER WIRE TO THE WHITE WIRES OF THE TWO LAMPS. PUT A LITTLE TAPE (ANY KIND) AROUND ONE OF THE THREE-WIRE SOLDER JOINTS, THEN TAPE BOTH SOLDER JOINTS TOGETHER SO THEY CAN'T TOUCH THE POWER JACK.

OPTIONAL

STRAIN RELIEF WIRES -- TO HELP PREVENT WIRES FROM FLEXING TOO MUCH DURING INSTALLATION AND BREAKING OFF WHERE THEY SOLDER TO THE BOARD. STRAIN RELIEF HOLES AND PADS HAVE BEEN PROVIDED ON THE LS//100A BOARD. THESE HOLES AND PADS ARE MARKED WITH A SMALL "s". CUT AND STRIP A SMALL PIECE OF WIRE SO THAT THE INSULATION IS AS LONG AS THE DISTANCE OVER THE WIRES AND BETWEEN THE TWO HOLES USED FOR STRAIN RELIEF. TIN ENDS OF WIRE AND PASS OVER WIRES AND DOWN THROUGH THE STRAIN RELIEF HOLES. PULL TIGHT, BEND OVER, AND SOLDER INTO PLACE.

NOTE: SOLDER WIRES TO POWER JACK AND NACELLE LEDs LATER (FOR A BOARD TEST TO BE SURE THERE ARE NO LOOSE WIRES OR UNSOLDERED I.C.s AND NACELLE PATTERN IS CORRECT) AFTER THE INSIDE OF THE MODEL HAS BEEN PAINTED BUT BEFORE COMPLETING THE ASSEMBLY OF THE MAJOR HULL PIECES. (ALSO, CHECK STEP 30)

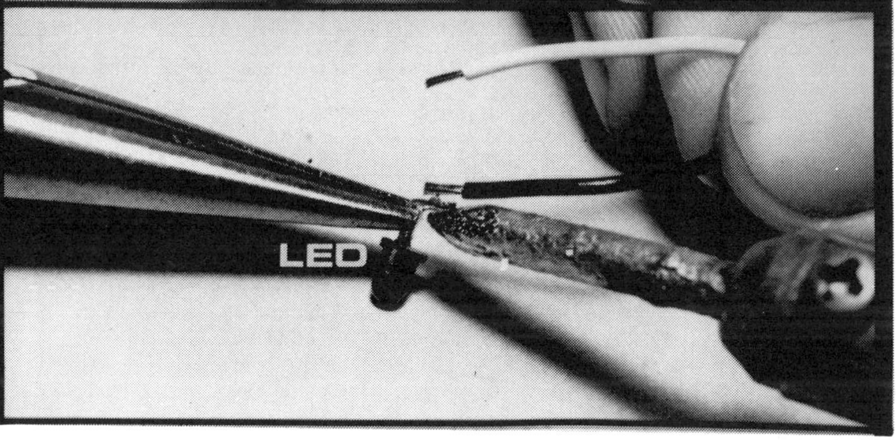

LS/100A Digital Electronics System NACELLE CABLE WIRING DIAGRAM

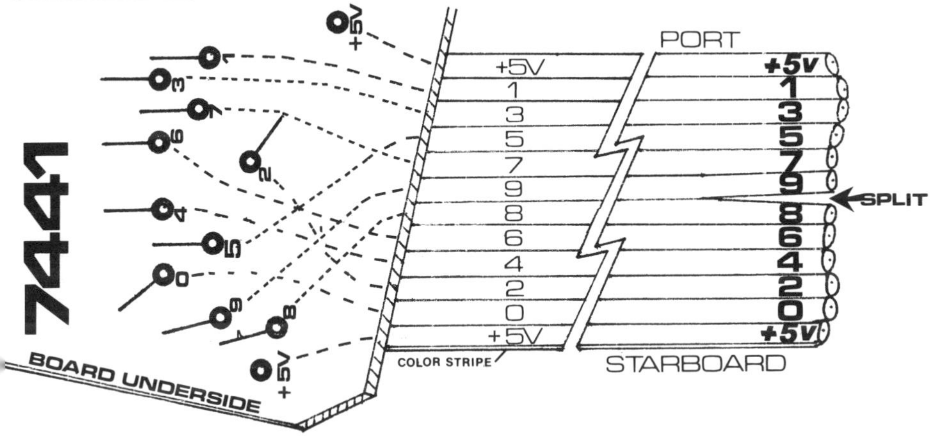

LS/100A
Digital Electronics System TM

STEP 4 - NACELLE DISC CONSTRUCTION

CUT TWO DISCS 7/8" IN DIAMETER FROM A SHEET OF .040" STYRENE. MARK WITH FIVE EQUALLY SPACED PENCIL LINES RADIATING FROM THE CENTER. CUT WITH A RAZOR SAW (GENTLY PULLING WITH THE TEETH OF THE SAW TOWARD YOU) TO WITHIN 5/32" FROM THE EXACT CENTER. HOLD THE DISC IN A VERTICAL POSITION. CAREFULLY WIDEN THE SLOTS UNTIL AN L.E.D. LEAD WILL FIT VERY SNUGLY INTO THE SLOT. INSERT 5 JUMBO CLEAR LEDs DOWN AGAINST THE DISC WITH THE FLATTENED RIM EDGE (NEGATIVE LEAD) POINTING OUT FROM THE CENTER. THE POSITIVE LEAD SHOULD BE SNUG AGAINST THE END OF THE SLOT TOWARDS THE CENTER OF THE DISC.

PUT A DROP OF SUPER GLUE ON THE UNDERSIDE OF THE LEDs - THIS SHOULD HOLD THEM, WHEN DRY, WHILE YOU CUT THE LED LEADS DOWN TO 3/8" LONG. BEND THE POSITIVE LEADS INTO THE CENTER UNTIL THEY TOUCH AT THEIR TIPS. SOLDER THEM TOGETHER WITH A SMALL BLOB OF SOLDER. THEN SOLDER THE POSITIVE 5 VOLT WIRE OF THE APPROPRIATE SECTION OF FLAT CABLE TO THE GROUPED POSITIVE LEADS.

NEXT, YOU SHOULD SOLDER THE REMAINING WIRES OF THE FLAT CABLE TO THE NEGATIVE LEADS OF THE NACELLES ACCORDING TO THE NACELLE L.E.D. WIRE DIAGRAM. IF YOU SOLDER THE NACELLE LEDs TO THE CABLE TO CHECK THE SYSTEM, YOU WILL HAVE TO DISCONNECT THESE CONNECTIONS TO STRAND THE WIRES THROUGH THE SECONDARY HULL PYLON SLOTS. SEE STEP #20D.

STEP 5 - CIRCUIT BOARD TESTING

TEST ENTIRE WORKING ELECTRONICS BY INSTALLING POWER PLUG AND JACK, WIRE CIRCUIT BOARD POWER LEADS TO JACK, THEN PLUG IN POWER AND TEST TO SEE IF EVERYTHING WORKS (RUNNING LIGHTS FLASH, NACELLE LIGHTS ROTATE IN PROPER SEQUENCE AND DIRECTION, AND INTERIOR LAMPS ALL COME ON) BEFORE PROCEEDING WITH CONSTRUCTION. IF ALL SYSTEMS WORK SUCCESSFULLY, THEN YOU HAVE COMPLETED THIS SECTION.

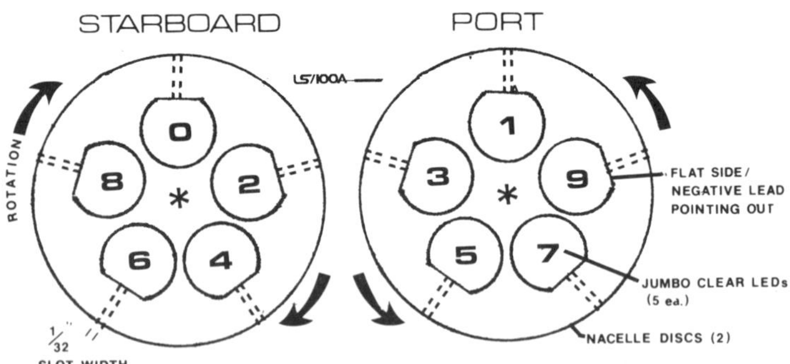

(NOT DRAWN TO SCALE) (REFER TO AP 03 11)

BOW PERSPECTIVE

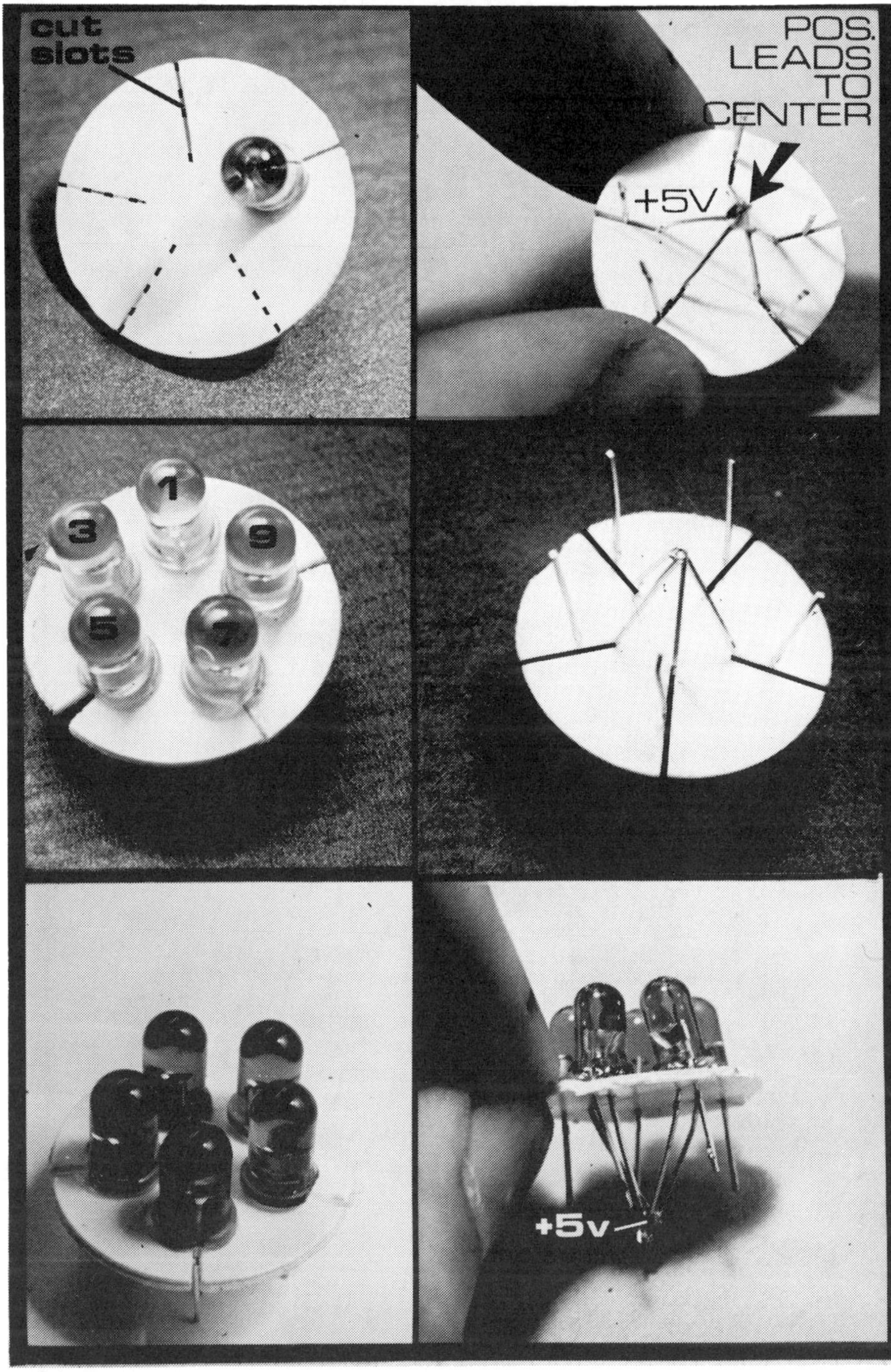
cut
slots
POS.
LEADS
TO
CENTER
+5V
1
3
9
5
7
+5v

TOOLS/MATERIALS

1A. TO COMPLETE THE HEAVY CRUISER STARSHIP AS EXPLAINED IN THIS MANUAL, YOU WILL NEED THE FOLLOWING TOOLS:

1 - FLUSH CUTTERS, 3/8" WIDE TEETH PREFERRED
1 - NAIL CLIPPERS
1 - NEEDLE-NOSE PLIERS
1 - RAZOR SAW, X-ACTO #235 (HANDLE OPTIÓNAL)
1 - FILE, METAL, MEDIUM TO LIGHT GRAIN
1 - NEEDLE FILE, BY X-ACTO #367
3 - PAINTBRUSHES, 00000, #4 MEDIUM
1 - FINE TIP FELT PEN, BLACK
1 - PR. SCISSORS
1 - X-ACTO KNIFE, #1, #11 BLADES
1 - HAND DRILL PIN VISE #7322
DRILL BITS:
#65 BIT IF USING 40 MIL FIBER OPTIC
#60 BIT IF USING 56 MIL FIBER OPTIC
#75 BIT IF USING 30 MIL FIBER OPTIC
1 - 1/8" DIAMETER BIT FOR GENERAL DRILLING
1 - 1/8" DIAMETER BIT FOR GENERAL DRILLING
1 - DEBURRER OR GRINDER BIT (DREMEL) #194, #953
1 - DREMEL TOOL (ANY MODEL)
OR VARIABLE SPEED HAND DRILL
1 - PACK FLEXIGRIT SAND PACKET, OR 3M SAND PACK #9055
5 - SMALL WIRE TWIST TIES

1B. THE FOLLOWING MATERIALS WILL BE REQUIRED TO COMPLETE YOUR HEAVY CRUISER STARSHIP USING THE LS//100A SYSTEM:

25 - 30 FT. 40 MIL FIBER OPTIC/INFO SHEET, SEE REFERENCES PAGE
1 - PLEXIGLAS ROD, 5/8" DIAMETER, 2-1/2" LONG
2 - NACELLE HALF-DOMES FROM ROMULAN KIT (OPTIONAL)
2 - PLASTIC CAPSULES FROM GUMBALL MACHINE
1 - PLASTRUCT PLASTIC STRIP #C-10, 5/16" X .080"
1 - PLASTRUCT TUBE #TB-10, 3/8" DIAMETER
1 - TUBE GREEN MAGIC PUTTY, OR TESTORS CONTOUR PUTTY
1 - TUBE "SUPER GLUE 3", OR "EASTMAN 910"
WARNING!! SHOULD BE USED WITH EXTREME CARE AND ADULT SUPERVISION.
1 - TUBE SET OF EPOXY GLUE
1 - ROLL MASKING TAPE, 3/4" WIDE
1 - ROLL SCOTCH TAPE (MAGIC MEND)
1 - BOTTLE "MICROWELD" PLASTIC ADHESIVE
1 - BOTTLE "MICRO-MASK" (OPTIONAL)
1 - ROLL BLACK MATTE GRAPHIC TAPE, 1/32" WIDE, FORMALINE, AVAILABLE AT ART OR GRAPHIC STORES

NOTES:

- ALSO CHECK WITH AP:03:04 FOR TOOLS AND MATERIALS OF THE LS//100A ELECTRONICS. SOME TOOLS AND MATERIALS MAY BE LISTED TWICE AND WILL REQUIRE SOME CROSS REFERENCE.
- ALSO CHECK THE MATERIALS REQUIRED FOR THE BASE ON PAGE AP:03:34.
- THE PAINT AND DECALS REQUIRED ARE LISTED ON PAGE AP:03:36 and :46.
- CHECK THE SUPERDETAILING SECTION FOR OPTIONS AND SUBSEQUENT MATERIALS.

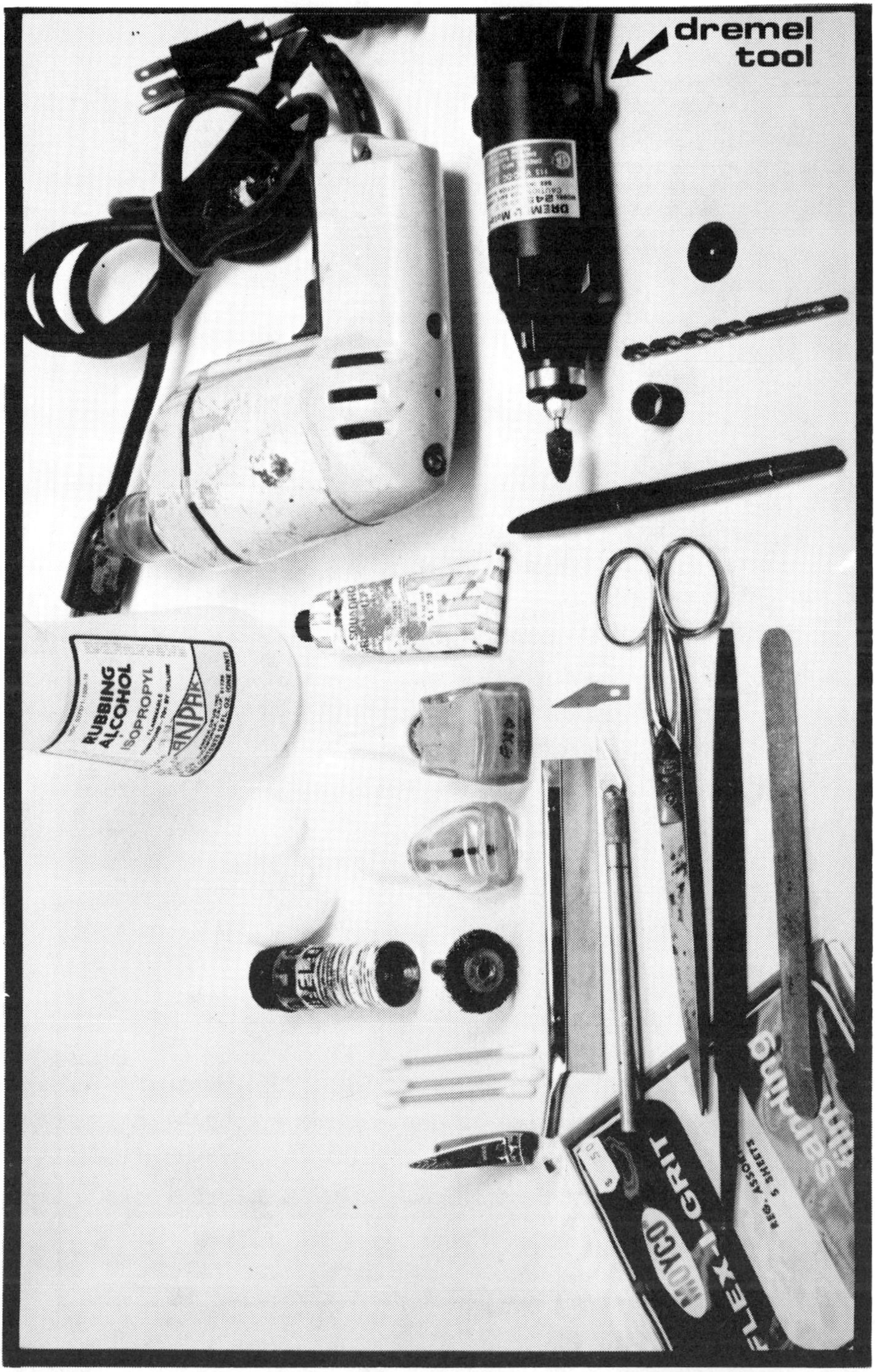
dremel
tool

PRE-ASSEMBLY

2 A. REMOVE ALL PARTS WITH AN X-ACTO KNIFE FROM THE SPRUES. LOCATE THE INSTRUCTIONS AND DECALS. THE GREEN SENSOR ARRAYS AND BASE WILL NOT BE NEEDED FOR THIS SYSTEM. CHECK WITH THE INSTRUCTION SHEET FOR ANY PARTS THAT MAY BE MISSING OR DAMAGED.

B. WITH COTTON OR A RAG SOAKED WITH ISOPROPYL (RUBBING) ALCOHOL, WIPE ALL SURFACES OF THE PLASTIC. THIS WILL REMOVE ANY OILS LEFT ON IT FROM THE STEEL MOLDS AT THE FACTORY.

C. SAND, TRIM OR FILE ALL ROUGH SPRUE MARKS AND GENERALLY SAND OVER ALL SURFACES. USE VERY FINE SANDPAPER.

D. <u>DRILL</u> <u>OUT</u> ALL PORTS AS LOCATED ON THE HULL PIECES BY RELIEFED CIRCLES AND RECTANGLES (PORTS). USE A <u>#65</u> <u>BIT</u> (IF USING 40 MIL FIBER OPTIC) AND A HAND TWIST DRILL.

E. <u>DRILL</u> <u>OUT</u> BOTH SENSOR ARRAYS IN BOTH HALVES OF THE PRIMARY HULL. THESE HOLES ARE TO BE 5/8" DIA. A SLIGHTLY SMALLER HOLE WILL GIVE SLIGHT TENSION TO THE PLEXIGLAS ROD (STEP #8a).

F. <u>DRILL</u> <u>OUT</u> A LONG HOLE LOCATED IN THE DORSAL CONNECTION OF THE LOWER HALF OF THE PRIMARY HULL. THIS WILL BE FOR THE WIRES OF THE LS//100A DIGITAL ELECTRONIC SYSTEM.

G. <u>DRILL</u> <u>OUT</u> A 1/4" DIA. HOLE ON THE BOTTOM SECTION OF THE SECONDARY HULL FOR THE BASE CONNECTION. LOCATE THIS HOLE ABOUT 3/4" BEHIND THE BASE HOLE ALREADY ON THIS PORT.SEE PAGE AP:03:34.

H. <u>DRILL</u> <u>OUT</u> FOUR 1/8" DIA. HOLES FOR RUNNING LIGHTS IN THE PRIMARY HULL. LOCATE THESE HOLES BY FOLLOWING THE HORIZONTAL LEFT AND RIGHT DEFLECTOR GRID LINES TO EACH RESPECTIVE SIDE. DRILL OUT THE PORT TOPSIDE HOLE 3/16" FROM THE EDGE OF THE HULL AND DIRECTLY CENTERED ON THE GRID LINE. DRILL OUT THE STARBOARD TOPSIDE HOLE IN THE SAME MANNER. DRILL A HOLE MEASURED EXACTLY 1/8" FROM THE OUTSIDE EDGE OF THE UNDERSIDE OF THE HULL, AND DIRECTLY BENEATH THE TOPSIDE RUNNING LIGHT WHEN VIEWED FROM THE SIDE. REPEAT FOR THE OTHER SIDE. LATER, WHEN THE RUNNING LIGHTS ARE IN PLACE, THE UNDERSIDE LIGHTS ARE TO BE SLIGHTLY CLOSER TO THE SENSOR ARRAY THAN THE TOPSIDE. ALL FOUR HOLES MAY REQUIRE TO BE "BEVELED" FOR CLEARANCE OF THE LEDs LATER.

I. <u>DRILL</u> <u>OUT</u> A 1/8" DIA. HOLE IN THE TOP OF THE SHUTTLE BAY DOORS. THIS IS FOR THE HOMING BEACON. IT MAY REQUIRE SOME MORE ENLARGEMENT LATER WHEN THE LED IS INSTALLED.

J. <u>DRILL</u> TWO 1/8" HOLES INTO THE RECTANGLES MARKED ON THE IMPULSE ENGINES. LATER SQUARE INTO RECTANGLES WITH AN X-ACTO KNIFE AND MINI FILE.

3 <u>SAND OFF THE FOLLOWING AREAS</u>:

A. THE DEFLECTOR GRID: TOPSIDE OF PRIMARY HULL. USE A FINE TO MEDIUM FINE SANDPAPER, CAREFULLY TO AVOID ANY DEEP SCRATCHES. USE AN X-ACTO KNIFE TO SCRAPE AT ANY TIGHT PLACES, SUCH AS THE BRIDGE SECTION OR DORSAL TOP-PLATE.

B. ALL RELIEFED PORTS, RECTANGLES AND CIRCLES LINES.

C. THE CIRCLES LOCATED ON THE UNDERSIDE OF THE PRIMARY HULL. AMT HAS 6 CIRCLES RELIEFED. THE 11 FT. MINIATURE HAD 3 CIRCLES <u>ETCHED</u> INTO THE HULL. USE A COMPASS OR SIMILAR INSTRUMENT TO ETCH 3 CIRCLES SPACED ROUGHLY THE SAME AS THE PREVIOUS 6.

4 <u>PUTTY THE THREE DEPRESSIONS WITHIN THE 6 CIRCLES ON THE PRIMARY HULL UNDERSIDE</u>.

NOTES:

- CHECK VARIOUS SOURCES FOR THE LOCATION OF PORTS. FILM CLIPS AND COLOR PHOTOS ARE MOST HELPFUL. ALTHOUGH THE PORTS ARE RECTANGULAR, THE ROUND FIBER OPTIC METHOD WORKS VERY WELL.

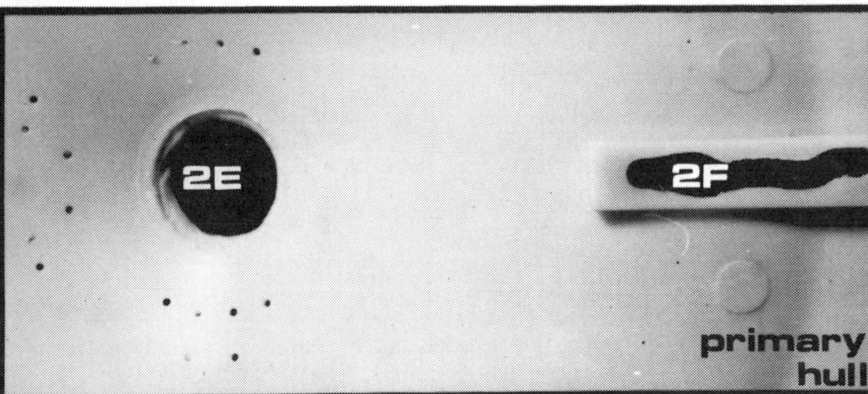
2E
2F
primary
hull

2D
2J

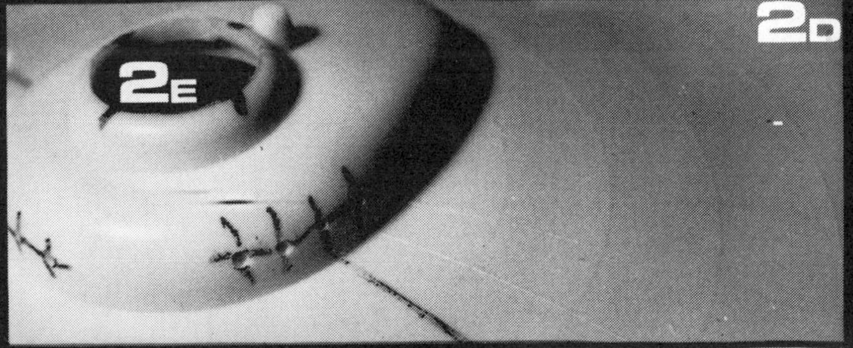
2D
2E

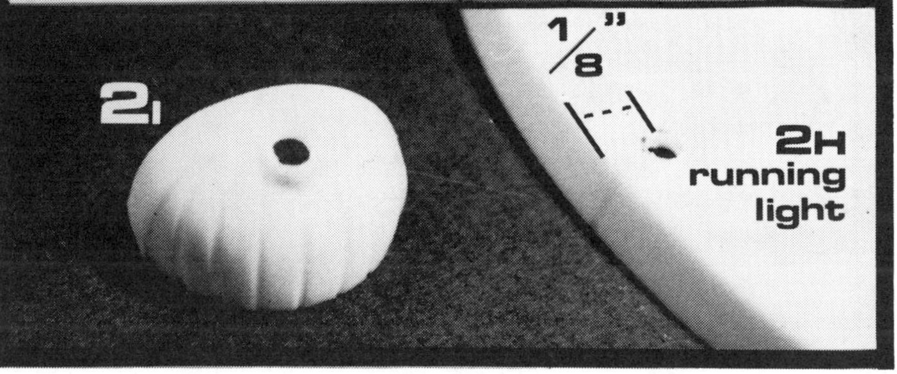
1/8"
2I
2H
running
light

5 PREPARE THE OTHER SMALLER PARTS OF THE SHIP

A. CUT THE SCREENS FROM THE INTERCOOLERS (OPTIONAL).

B. SAND SEAMS AND PUTTY MOLD MARKS ON ALL INTERCOOLERS AND REACTORS.

C. TRIM 1/8" TO 1/4" OFF OF THE ANTENNA MOUNTING STEM.

D. TRIM TWO HALF DOMES FROM THE ROMULAN KIT TO FIT ON THE REAR NACELLE CAPS.

E. FINISH SQUARING OFF THE RECTANGLES IN THE IMPULSE ENGINES

F. WITH A RAZOR SAW, CUT THE DOMES FROM THE FRONT NACELLE CAPS, LEAVING TWO RINGS. USE A GENTLE PULLING ACTION WHEN CUTTING WITH THE TEETH OF THE BLADE FACING YOU.

G. TAKE TWO CLEAR PLASTIC CAPSULES (FROM A GUMBALL MACHINE) AND CUT TWO DOMES TO REPLACE THOSE PREVIOUSLY CUT FROM THE NACELLE CAPS. APPROXIMATE THE DOME DIMENSION BY CONSULTING PHOTOS AND BLUEPRINTS. SAND DOMES WITH VERY FINE SANDPAPER TO ACHIEVE A FROSTED EFFECT. REMEMBER TO USE A GENTLE PULLING ACTION DUE TO THE FRAGILE PLASTIC, WHICH CRACKS EASILY.

6 A. PAINT ALL PARTS (EXCEPT CLEAR DOMES) FLAT BLACK. SOME PAINT MAY CLOG IN THE PORTS, HOWEVER, THEY CAN BE CLEARED WITH A PIN. AFTER THE FLAT BLACK HAS DRIED, PAINT ALL INSIDE HULL SURFACES BRIGHT SILVER TO REFLECT THE LIGHT SOURCE FOR THE FIBER OPTIC.

B. SCRAPE PAINT OFF OF ALL SURFACES TO BE GLUED.

7 PREPARE A PIECE OF PLEXIGLAS ROD MEASURING 5/8" DIA. X 2-1/16" LONG.

A. WITH A METAL FILE, GRIND DOWN THE ROD TO A GENTLE SLOPE AT EACH END. ROTATE THE ROD WHILE FILING TO OBTAIN AN EVEN SURFACE. EACH END IS IDENTICAL IN SHAPE. SAND WITH SANDPAPER TO A FROSTED SURFACE.

B. AT ONE END, DRILL FOUR 1/8" HOLES INTO THE ROD AT AN ANGLE.

C. AT THE SAME END, WRAP 1/32" GRAPHIC TAPE SEVERAL TIMES AROUND THE ROD TO ACT AS A SPACER RING BEFORE PLACING INTO HULL SECTION.

NOTES:

- REFER TO THE SUPERDETAILING SECTION FOR FURTHER MODIFICATIONS THAT MAY REQUIRE ATTENTION AT THIS POINT.
- THE INTERCOOLERS ON THE 11 FT. STUDIO MINIATURE HAS SCREENS, HOWEVER, SOME SCENES IN THE EPISODES USING THE 4 FT. MINIATURE SHOWED NO SCREENS. THE HEAVY CRUISER BLUEPRINTS SHOW NO SCREENS ON THE INTERCOOLERS.
- THE DOMES OFF OF THE ROMULAN KIT ARE SLIGHTLY LARGE, BUT ARE A GREAT IMPROVEMENT OVER THE AMT VERSION.
- NOT ALL GUMBALL MACHINE CAPSULES ARE IDENTICAL.
- CHECK THE HEAVY CRUISER BLUEPRINTS (FRANZ JOSEPH) TO SHAPE THE SLOPE OF THE SENSOR ARRAYS (SHEETS 1,4, and 5).
- LEAVE THE ANGLED "LIP" ON THE REMAINING RING WHILE CUTTING ON THE ORIGINAL NACELLE CAP.
- PAINTING THE PARTS FLAT BLACK WILL BLOCK THE INTERIOR LIGHT FROM PENETRATING THE PLASTIC. TO MAKE SURE, HOLD THE PAINTED PIECES UP TO A BRIGHT LIGHT TO CHECK FOR ANY UNEVEN SPOTS.

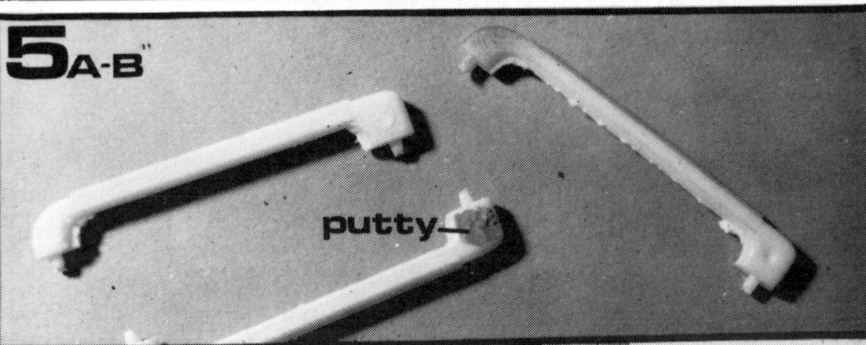
5A-B"
putty

5G
frost

5F
trim
cut

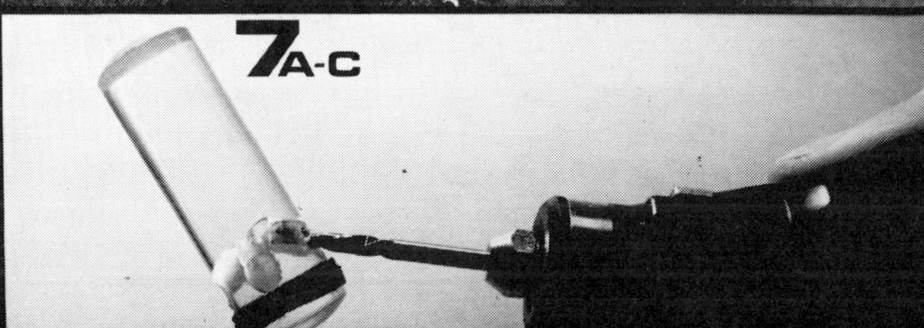
7A-C

PRIMARY HULL

8 TAKE THE COMPLETELY DRILLED AND PRIMERED LOWER SECTION OF THE PRIMARY HULL AND COMPLETE THE FOLLOWING:

A. CEMENT THE PREPARED PLEXIGLAS ROD INTO PLACE WITH 1/32" EXCESS OUT OF THE BOTTOM (AS SHOWN) WITH THE END WITH DRILLED HOLES FACING DOWNWARD.

B. WHEN THE ROD HAS DRIED, WRAP 1/32" WIDE GRAPHIC TAPE AROUND THE EXCESS TO MAKE A RING. THIS WILL COVER OVER ANY MISTAKES FROM DRILLING THE BOTTOM SENSOR ARRAY HOLE.

9 STRAND 40 MIL (OR A DIAMETER OF YOUR CHOICE) FIBER OPTIC THROUGH THE PORT HOLES LOCATED NEAR THE BOTTOM SENSOR ARRAY AND LOCATE THE ENDS INTO THE HOLES IN THE PLEXIGLAS ROD, USING NEEDLE-NOSE PLIERS TO POSITION THE FIBER OPTIC (AS SHOWN). LEAVE 1/4" EXCESS PROTRUDING THROUGH THE PORT (OUTSIDE); GLUE WITH A DROP OF SUPER GLUE ON THE INSIDE OF THE HULL AT EACH PORT.

10 PLACE THE COMPLETED LS//100A CIRCUIT BOARD ONTO THE PLEXIGLAS ROD BUT DO NOT CEMENT THE BOARD INTO PLACE YET.

A. LOCATE THE FOUR LEDs TO THEIR RESPECTIVE PLACES: TOPSIDE, PORT - RED, STARBOARD - GREEN, UNDERSIDE, PORT AND STARBOARD - YELLOW. DO NOT CEMENT.

B. RUN THE FOLLOWING WIRES DOWN THE DORSAL CONNECTION:
   1. NACELLES "CABLE"
   2. HOMING BEACON WIRE
   3. POWER CONNECTION WIRE
   4. FILAMENT BULBS (3) AND WIRE LENGTHS

11 PLACE 1/2" STRANDS OF FIBER OPTIC INTO THE PORT HOLES OF THE TOP SECTION OF THE PRIMARY HULL.

NOTES:

- WHEN GLUING THE PLEXIGLAS ROD INTO PLACE, USE A VERY QUICK SETTING AND STRONG GLUE LIKE "SUPER GLUE 3" OR "EASTMAN 910". PLEASE BE EXTREMELY CAREFUL WHEN USING THESE SUPER GLUES!
- THE EXCESS OF PLEXIGLAS ROD ON THE UNDERSIDE OF THE PRIMARY HULL IS FOR A RING MADE FROM GRAPHIC TAPE. THIS METHOD NOT ONLY COVERS OVER ANY MISTAKES, BUT DUPLICATES THE DETAIL ON THE ACTUAL STARSHIP ENTERPRISE.
- THE FIBER OPTIC PLACED IN THE TOP HALF OF THE PRIMARY HULL WILL MAKE CONTACT WITH THE PLEXIGLAS ROD WHEN THE TOP IS PLACED INTO POSITION, AND TRANSFER LIGHT TO THE PORTS WHEN THE ROD IS LIGHTED.
- THE EXCESS FIBER OPTIC TO PROTRUDE THROUGH ALL PARTS WILL BE TRIMMED OFF WITH FLUSH CUTTERS AFTER THE SHIP HAS BEEN PAINTED.
- IF NOT USING THE LS//100A KIT, USE SUB-MINIATURE LEDs FOR THE RUNNING LIGHTS. SMALL FILAMENT BULBS WILL NOT WORK WITH THIS TYPE OF SYSTEM.

SOME WORDS ABOUT FIBER OPTIC:

FIBER OPTIC IS A RELATIVELY NEW MATERIAL DESIGNED TO TRANSMIT LIGHT NEARLY EQUAL TO THAT OF ITS SOURCE. IT LOOKS AND HAS CHARACTERISTICS VERY MUCH LIKE THAT OF HEAVY FISHLINE. THE LARGER THE DIAMETER, THE MORE EVEN THE LIGHT APPEARS AT THE END, AND IN OUR CASE IT REPRESENTS THE DIFFUSION OF LIGHT THROUGH A PORT MORE REALISTICALLY THAN JUST A SINGLE LIGHT INSIDE A PLASTIC SHELL WITH HOLES. SHIPS IN PRODUCTIONS LIKE BATTLESTAR GALACTICA, STAR WARS, AND CE3K USE FIBER OPTIC BECAUSE OF THE INDIVIDUAL INTENSITY OF EACH PORT'S LIGHT.

WORKING WITH FIBER OPTIC IS NOT TOO DIFFICULT. TO MAKE AN EVEN CUT, NECESSARY FOR THE BEST ANGLE OF LIGHT ACQUISITION, USE SOME SIMPLE NAIL-CUTTERS. FIBER OPTIC CAN BE BENT AT NEAR 90° ANGLES CAREFULLY WITH NEEDLE-NOSE PLIERS. SOME PRACTICE MAY BE NECESSARY BEFORE ATTEMPTING TO INSTALL IT INTO THE SHIP.

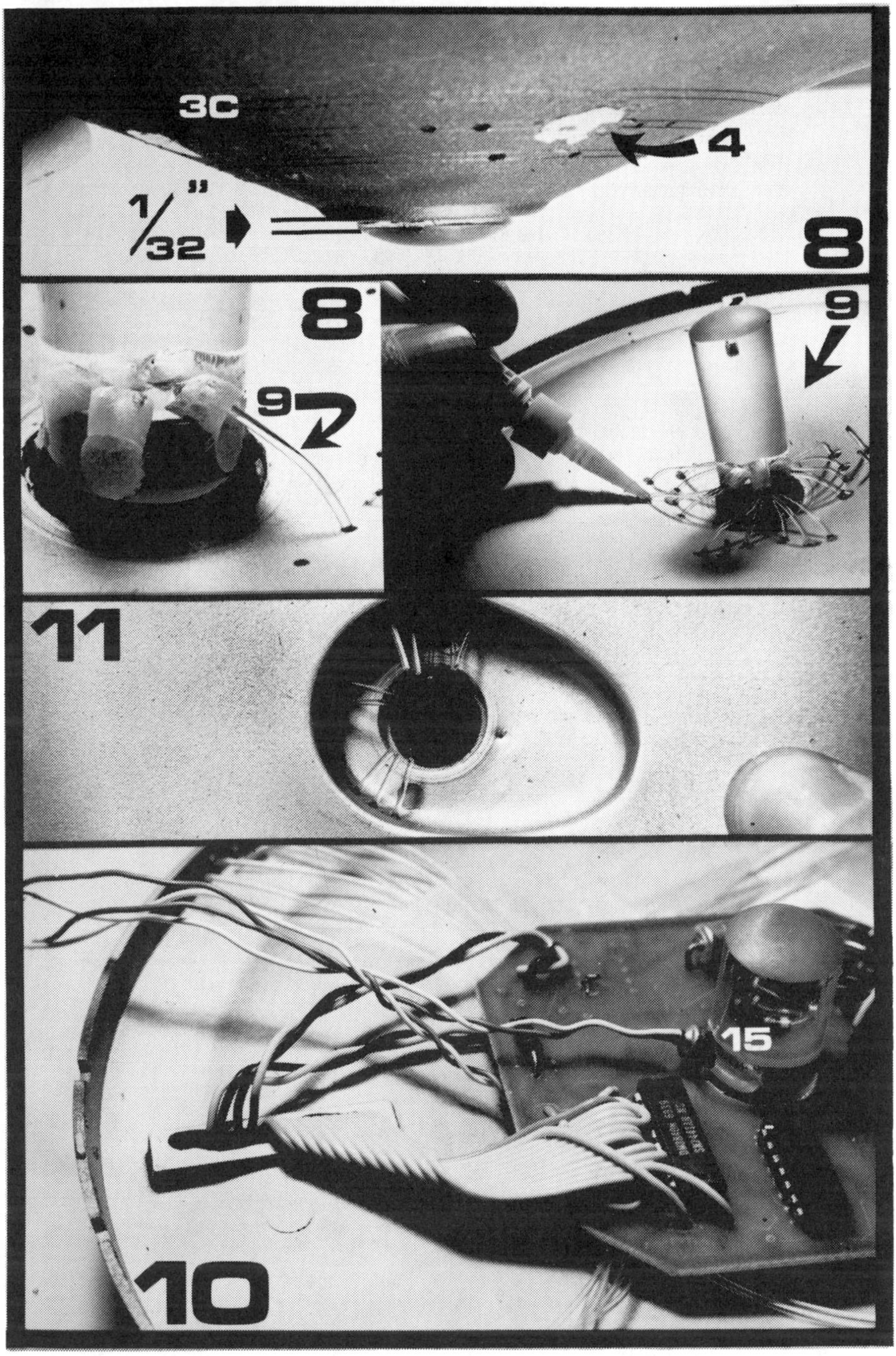
3C
4
1/32"
8
8
9
9
11
15
10

12 TAKE A PIECE OF PLASTRUCT TUBING (TB-10) AND CUT A PIECE ABOUT 1-1/4" IN LENGTH. DRILL A HOLE 3/16" DIA. INTO THE CENTER SIDE OF THE TUBING AND THEN PLACE A SMALL FILAMENT LAMP (RADIO SHACK #272-1140, 6V, 25 ma) INTO THE HOLE. THIS WILL SERVE AS A LIGHT SOURCE. GLUE INTO PLACE (AS SHOWN) AT A LOCATION JUST BELOW THE EDGE OF THE CIRCUIT BOARD. USE EPOXY GLUE AND SCRAPE AWAY PAINT FROM THE HULL WHERE CONTACT IS TO BE MADE. THIS MAY REQUIRE LIFTING OF THE CIRCUIT BOARD TO POSITION THE TUBE INTO PLACE. SINCE THE HULL HAS SOME CURVATURE, ALSO TAPE INTO PLACE WHEN GLUING.

13 STRAND MORE FIBER OPTIC THROUGH THE GROUPS OF PORTS AT THE EDGE OF THE HULL. RUN ALL STRANDS ON THE PORT SIDE INTO THE LEFT END OF THE TUBE. LIKEWISE FOR THE STARBOARD/RIGHT SIDE - LEAVE THE FILAMENT LAMP IN PLACE WHEN STRANDING FIBER OPTIC INTO THE TUBE.

14 WHEN COMPLETED WITH THE FIBER OPTIC, AGAIN LOCATE THE FOUR RUNNING LIGHTS, AND BEVEL OUT THE INSIDE EDGES OF THE HOLES TO ALLOW THE LIGHTS TO PROTRUDE THE MAXIMUM HEIGHT. SOME LEDs MAY BE HIGHER THAN OTHERS. USE A UNIFORM AMOUNT OF DISTANCE (AS SHOWN). GLUE INTO POSITION WITH SUPER GLUE ON THE UNDERSIDE, THEN PLACE TAPE OVER EACH SET OF LED LEADS SO THAT THERE IS NO SHORTING OUT OF LEDs WHEN THE TOP AND BOTTOM SECTIONS ARE JOINED.

15 AT A LOCATION ON THE PLEXIGLAS ROD, JUST ABOVE THE CIRCUIT BOARD (SEE PHOTO WITH "X" MARKED ON ROD), DRILL OUT A HOLE EQUAL TO THAT OF ANOTHER 6V FILAMENT LAMP. DRILL ABOUT 3/4 OF THE WAY INTO THE ROD SO THAT THE FILAMENT OF THE BULB IS EXACTLY IN THE CENTER OF THE ROD.

16 NOW TEST THE CIRCUIT BOARD AND FILAMENT BULB CONNECTION. REMEMBER THAT IT HAS ONLY ONE POLARITY. IF THERE IS NO PROBLEM, AND THE RUNNING LIGHTS (AND HOMING BEACON AND FILAMENT LAMPS) WORK PROPERLY, YOU CAN GLUE THE BOARD INTO PLACE, BUT DO NOT GLUE THE TOP SECTION OF THE PRIMARY HULL YET INTO PLACE.

NOTES:

- NOT ALL PORTS OF THE SHIP ARE LIGHTED AT ANY ONE TIME, SO SOME PORTS CAN BE INSERTED WITH SMALL PIECES OF FIBER OPTIC, AND NOT AIMED AT THE LIGHT SOURCE. HOWEVER, 30 FT. IS THE EXACT AMOUNT REQUIRED IF YOU WISH TO LIGHT ALL PORTS.
- 40 MIL FIBER OPTIC IS SUGGESTED FOR THIS MODEL KIT. 56 MIL IS TOO DIFFICULT TO BEND IN TIGHT SPACES. SMALLER DIAMETERS (I.E., 30, 20, 10 MIL) **DO NOT APPEAR AS REALISTIC PORTS, BUT RATHER PIN-POINTS OF LIGHT.**

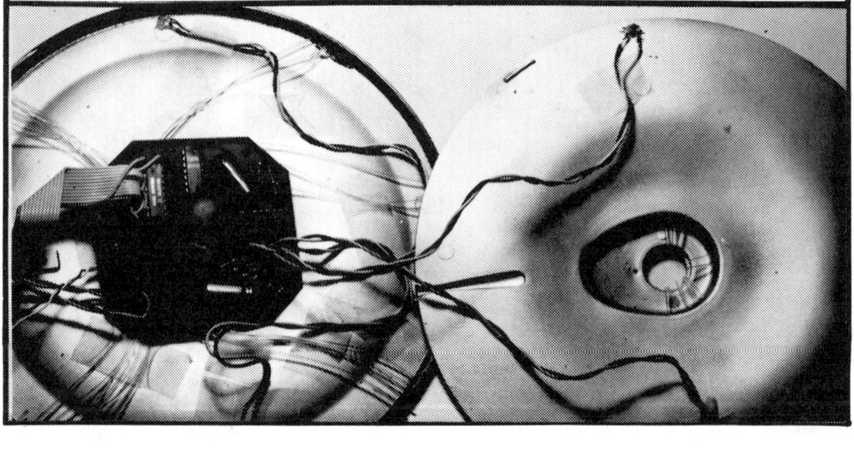

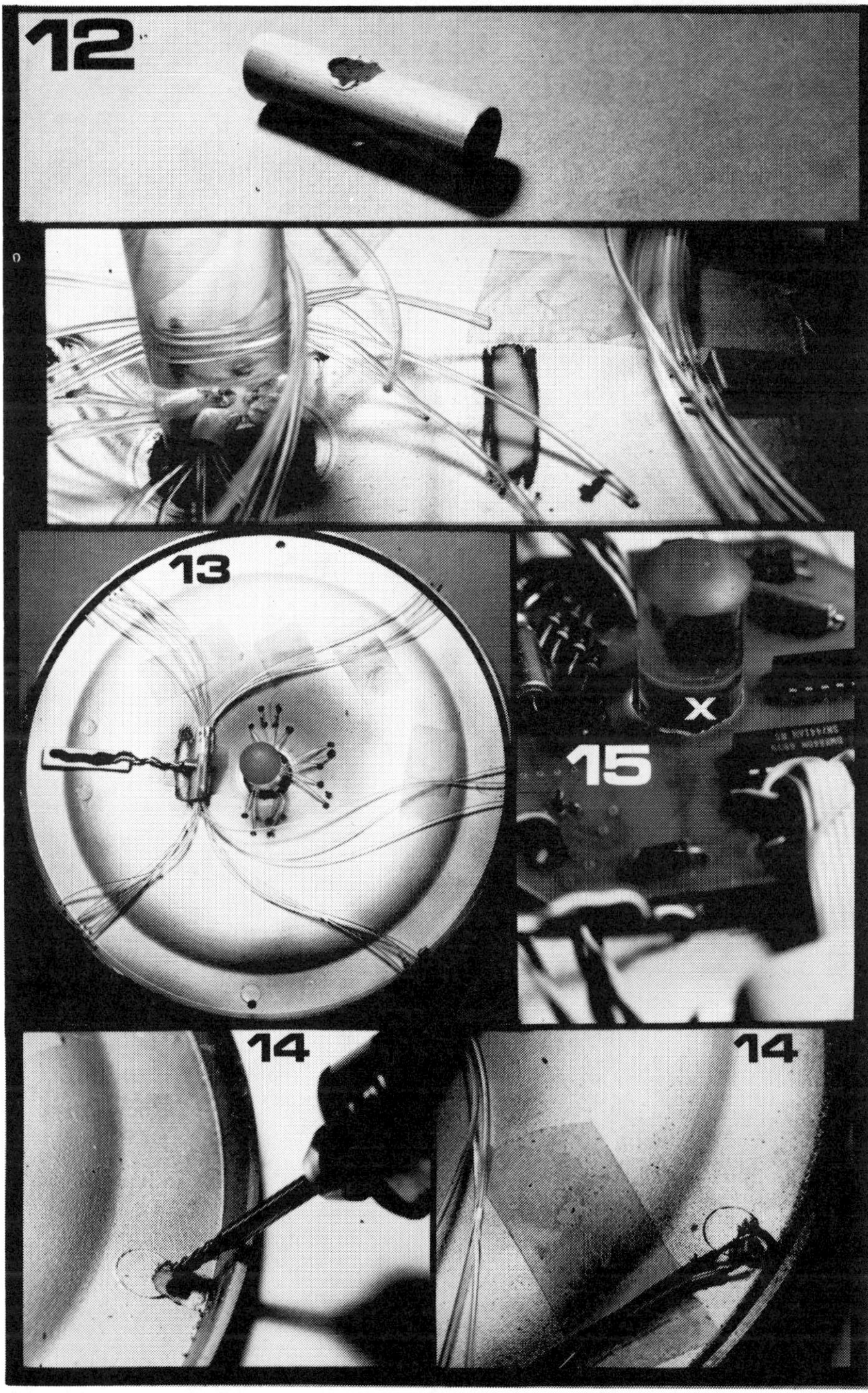
12
13
X
15
14
14

DORSAL

17 TAKE THE TWO TOP HALVES OF THE SECONDARY HULL WITH THE DORSAL (OR VENTRAL FIN) ATTACHED AND COMPLETE THE FOLLOWING:

A. INSERT MORE FIBER OPTIC STRANDS (AS SHOWN). START WITH THE DORSAL SECTIONS, ADDING A DROP OF SUPER GLUE TO EACH STRAND TO HOLD IT IN PLACE. REMEMBER TO LEAVE AN EXCESS PROTRUDING THROUGH THE OUTSIDE OF THE HULL.

B. NOW INSERT FIBER OPTIC INTO THE TOP SECTIONS OF THE SECONDARY HULL.

C. WHEN ALL FIBER OPTIC HAS BEEN INSERTED AND GLUED INTO PLACE, GROUP THE STRANDS WITH A TWIST-TIE TIGHTLY AND WITHOUT BREAKING ANY FIBER OPTIC. THIS IS A DIFFICULT PROCEDURE, AND SHOULD BE DONE CAREFULLY.

D. WITH NAIL CLIPPERS OR FLUSH CUTTERS, TRIM OFF ALL THE STRANDS THAT ARE BUNDLED TO A FLAT EDGE.

E. CUT ANOTHER PIECE OF PLASTRUCT TUBING (TB-10) TO ABOUT 3/4" TO 1" AND CAREFULLY PLACE OVER THE BUNDLED FIBER OPTIC, ABOUT HALF-WAY INTO THE TUBE AND GLUE, ALLOWING ROOM FOR THE REMAINING TWO FILAMENT BULBS TO BE INSERTED IN THE OTHER END.

F. REPEAT FOR OTHER HALF.

G. REMEMBER TO POSITION THE LIGHT SOURCE TUBES AFTER THE TWO HALVES HAVE BEEN JOINED. THEREFORE, DO NOT GLUE THE TUBES INTO PLACE ON THE INSIDE HULL SURFACE YET.

NOTES:

- THIS PROCEDURE IS SOMEWHAT DIFFICULT AND CARE AND PATIENCE IS REQUIRED. MAKE SURE THAT THE ENDS OF THE FIBER OPTIC ARE BENT AT 90° NEAR THE END BEFORE INSERTING INTO THE PORTS.

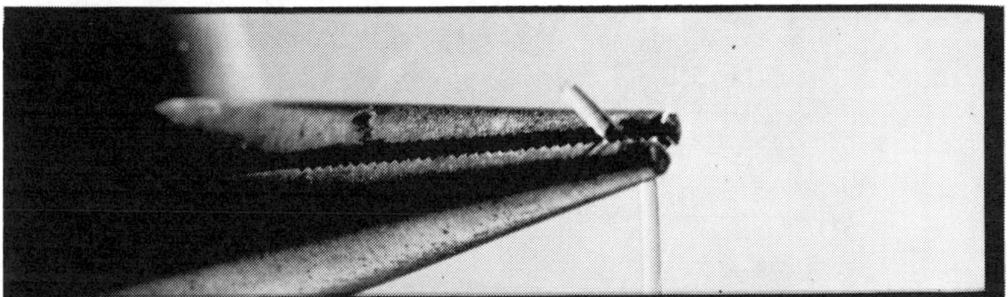

17
20c
grind

17B-F
19

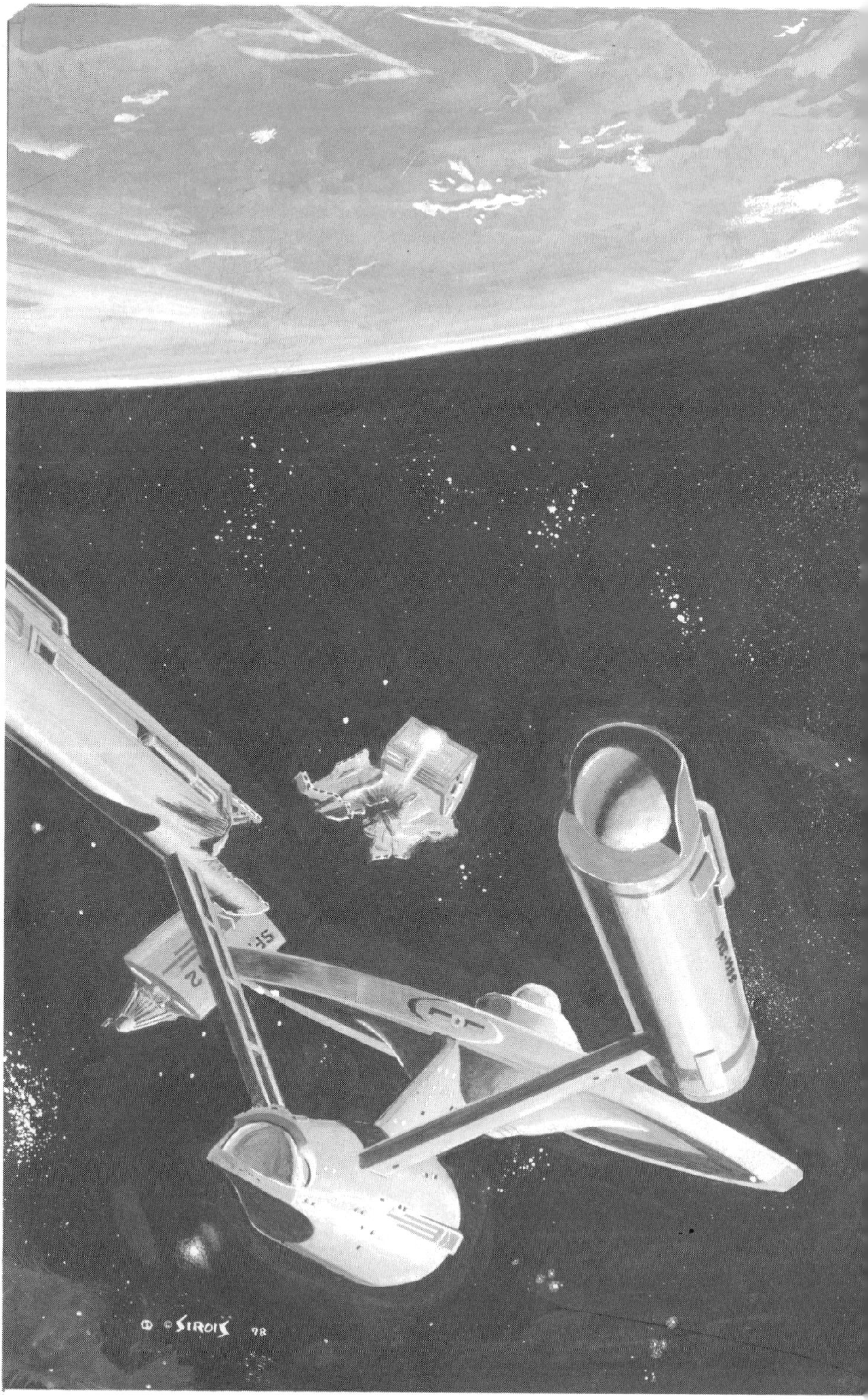
© SIROIS 78

SFD
A70
B21
SFD-A70

SECONDARY HULL

18 COMPLETE THE FOLLOWING ON THE BOTTOM HALF OF THE SECONDARY HULL:

A. TAKE THE PHONE JACK (RADIO SHACK #274-252) AND MOUNT INTO THE 1/4" DIA. HOLE DRILLED IN THE HULL. IF THE HOLE HAS BEEN DRILLED AT SLIGHTLY LESS THAN 1/4", THE PHONE JACK CAN BE TWISTED INTO PLACE, WITH THE NUT AND WASHER ACTING AS A SPACER RING. GLUE INTO POSITION WITH SUPER GLUE. (SEE NOTES BELOW).

B. TAKE SEVERAL PRE-CUT PIECES OF FIBER OPTIC OF ABOUT 1" IN LENGTH AND INSERT INTO THE SIDE PORTS OF THIS HULL SECTION. THEN, WITH NEEDLE-NOSE PLIERS, BEND EACH STRAND AT A 90° UPWARD ANGLE.

C. GLUE THE SHUTTLE DOORS INTO PLACE. THE BOTTOM CORNERS OF THE DOORS MAY BLOCK THE SIDE REAR PORTS IN THE HULL SECTION. DRILL OUT IF SO.

19 ON THE TOP HALVES OF THE SECONDARY HULL, THERE ARE TWO SUPPORT PYLON SLOTS. DRILL OUT THE BOTTOM OF EACH SLOT TO ALLOW THE NACELLE CABLE TO STRAND THROUGH. SPLIT THE CABLE EXACTLY IN HALF, WITH 6 STRANDS IN EACH HALF.

20 CAREFULLY TAKE BOTH SECONDARY HULL TOP SECTIONS, AND COMPLETE THE FOLLOWING:

A. WITH ONE SECONDARY HULL SECTION, TEST FIT INTO THE PRIMARY HULL DORSAL CONNECTION. YOU WILL BE REQUIRED TO DRILL OR GRIND OUT THE TOP EDGE OF THE INSERT OF EACH HALF. DO NOT CUT OFF THE INSERTS OF THE PYLONS, JUST THE RAISED EDGE THAT WOULD BLOCK THE WIRES.

B. STRAND ALL WIRES FROM THE PRIMARY HULL DOWN ALONG THE INSIDE EDGE OF THIS DORSAL HALF. THIS IS DIFFICULT AND WILL REQUIRE SOME POSITIONING OF WIRES AND FIBER OPTIC, SO TAKE YOUR TIME. IF ALL WIRES AND FIBER OPTIC ARE IN PLACE, THEN GLUE THIS HALF IN POSITION WITH SUPER GLUE AT THE PRIMARY HULL CONNECTION.

C. TAKE THE OTHER HALF AND DRILL OUT THE SAME AREA (AS SHOWN) AND POSITION INTO PLACE. THIS STEP IS CRITICAL AND SOMEWHAT DIFFICULT. FIRST GLUE AT THE PRIMARY HULL CONNECTION, THEN DOWN THE LEADING SIDES OF EACH DORSAL HALF, WITHOUT GETTING ANY GLUE ON YOUR HANDS. USE TAPE TO HOLD THESE TWO HALVES IN PLACE WHILE GLUING. ALSO, REMEMBER TO SCRAPE ALL PAINT OFF ANY SURFACES TO BE GLUED, AND THAT HEAVY PRESSURE MUST BE APPLIED TO MAKE THE TIGHTEST BOND WHEN USING SUPER GLUE. IT WILL NOT HOLD IN GAPS IN THE POINTS OF CONTACT.

D. KEEP THE TAPE ON THE DORSAL SECTION AND AFTER SUFFICIENT DRYING TIME, STRAND THE TWO GROUPS OF NACELLE CABLE THROUGH THE PRE-DRILLED PYLON SLOTS (AS SHOWN). EITHER SET OF WIRES CAN BE STRANDED IN EITHER SLOT.

E. THE SEAM RUNNING DOWN THE TOP OF THE SECONDARY HULL WILL NOT BE GLUED YET.

NOTES:

- IF POSSIBLE, TWIST THE PHONE JACK INTO THE 1/4" HOLE UNTIL IT MAKES A FLUSH FIT ON THE BOTTOM OF THE HULL.

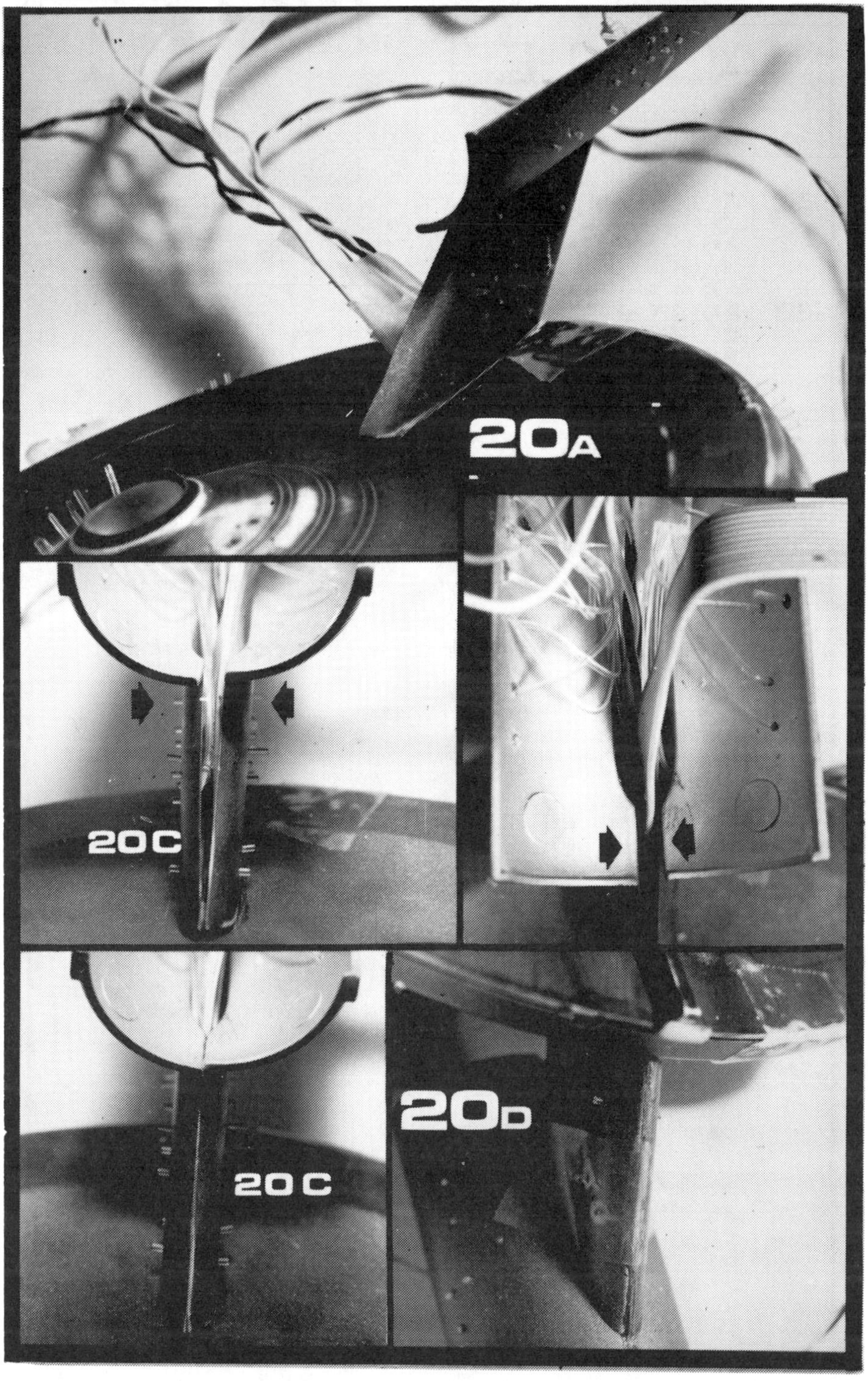
20A
20C
20C
20D

NACELLES

21 PREPARE THE NACELLES/PYLONS SECTIONS WITH A DEBURRER BIT (DREMEL TOOL). GRIND DOWN THE RELIEFED EDGE AT THE "TONGUE" OF EACH PYLON (AS SHOWN) SO THAT IT WILL CLEAR THE NACELLE WIRES.

22 WITH THE NACELLE WIRES EXTENDING FROM THE PYLON SLOTS, POSITION ONE HALF OF THE NACELLE SECTIONS INTO THE PYLON SLOT AND POSITION THE WIRES UP THE PYLON. DIVIDE THE WIRES AND ROUTE AROUND THE PIN IN THE CENTER OF THE PYLON. CONTINUE TO ROUTE THE WIRES UP TO THE FRONT OF THE NACELLE.

23 REPEAT FOR OTHER NACELLE.

24 ASSEMBLE THE NACELLE DISC WITH LEDs (PRE-SOLDERED AND TINNED) INTO THE NACELLE RING YOU FABRICATED IN STARSHIP PRE-ASSEMBLY. PUSH THE NACELLE DISC ALL THE WAY INTO THE RING, SO THAT IT STOPS AT THE LIPPED EDGE OF THE RING.

25 REFER TO THE SECTION ON NACELLE WIRING WITH THE DIAGRAM ON PAGE AP:03:11 TO WIRE EACH SET OF LEDs ACCURATELY. ONE POWER WIRE GOES TO THE CENTER AND 5 GROUND WIRES GO TO THE 5 OUTSIDE LEADS. IF THE WIRING DIAGRAM IS CONFUSING OR TOO DIFFICULT, ATTACH ONE WIRE AT A TIME (WITH THE SYSTEM ON); ONCE THE CENTER WIRE HAS BEEN SOLDERED, THEN VISUALLY DECIDE UPON WHICH WIRE IS WHICH, BY TOUCHING EACH WIRE RANDOMLY TO A LED LEAD. THIS IS NOT AS DIFFICULT TO DETERMINE AS YOU MIGHT THINK. ALSO CHECK PAGE AP:03:10 FOR HINTS ON SOLDERING. AVOID PROLONGED HEAT FROM SOLDERING IRON ON LEDs.

26 IMPORTANT NOTE: THE NACELLE WARP ACQUISITION FANS ROTATE TOWARDS THE CENTER OF THE SHIP. THAT IS, AS VIEWED FROM THE BOW, THE STARBOARD NACELLE FAN ROTATES CLOCKWISE, AND THE PORT NACELLE FAN ROTATES COUNTER-CLOCKWISE. A RATE OF ABOUT FOUR REVOLUTIONS/SEC IS NORMAL (THIS RATE HAS BEEN PRESET BY THE LS//100A KIT). BOTH NACELLE FANS ARE SYNCHRONIZED AS BOTH LEDs NEAREST TO EACH OTHER ARE ON, AND THOSE FARTHEST APART ARE ON AT THE SAME TIME. SO, BEFORE GLUING THE RING/LED ASSEMBLY INTO PLACE, VISUALLY ALIGN THE SYNCHRONIZATION BY TURNING ONE OF THE RINGS INTO ITS PROPER POSITION. SEE PAGE AP:03:12.

27 NOW BRING THE OTHER HALF OF THE NACELLE SECTIONS INTO PLACE AND CAREFULLY ALIGN THE SEAMS TO AVOID ANY OVERHANG AND UNNECESSARY FILING LATER. SOME NACELLE SECTIONS COULD BE IMPROPERLY MOLDED AND HAVE A TERRIBLE "TWIST" TO THEM. THESE PARTS WILL REQUIRE SOME CAREFUL ALIGNMENT AND THE USE OF TAPE FOR PROPER ALIGNMENT. (REFER TO SFAM #1, AP:01:18 FOR MORE DETAILS).

28 ONCE THE NACELLES ARE IN PLACE, SAND AND FILE ALL SEAMS. ESPECIALLY FILE THE LOUVERED RINGS - (JUST BEHIND THE NACELLE RING SECTION YOU JUST GLUED ON). IF THE SEAMS WERE NOT POSITIONED CORRECTLY, USE A CROSSCUTTING ACTION WHEN FILING THE SEAMS).

NOTES:

- COAT EACH LED CONNECTION WITH A HEAVY PAINT OR POLISH TO PREVENT ANY "SHORTING OUT" OF LEDs.
- PROLONGED HEAT ON LEDs WILL BURN OUT THE TINY CIRCUITRY.

21
grind

24
nacelle ring

22
split

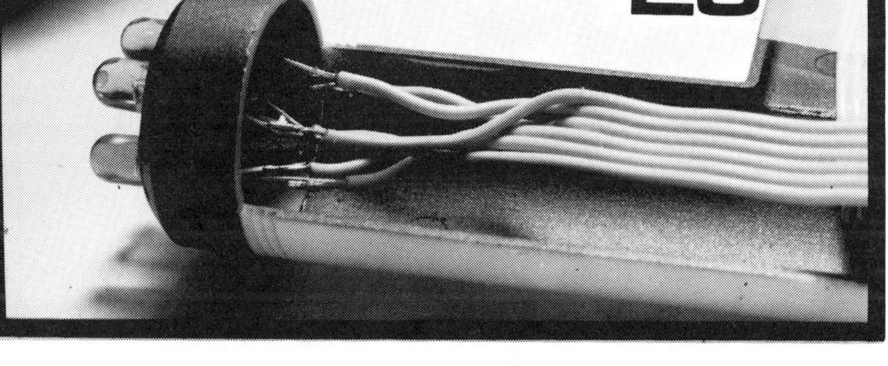
25

FINAL ASSEMBLY

29 NOW GLUE THE TOP SEAM OF THE SECONDARY HULL. WORKING WITH THE NACELLES/PYLONS WOULD HAVE CAUSED CONSTANT STRESS ON THIS POINT AND REPEATED BREAKAGE. CAREFULLY ALIGN AND GLUE. THE BETTER THE ALIGNMENT, THE LESS SANDING WILL BE REQUIRED LATER.

30 CONNECT THE POWER WIRES TO THE PHONE JACK AS ILLUSTRATED AND SOLDER. TO ASSURE THE PROPER CONNECTION, ASSEMBLE THE BASE AS DESCRIBED ON PAGE AP:03:34 BEFORE CONTINUING, OR AT LEAST WIRE UP A PHONO PLUG TO DETERMINE POLARITY. IF THE FILAMENT LIGHTS WORK, BUT THE NACELLES AND RUNNING LIGHTS DO NOT, THEN THE POLARITY HAS BEEN REVERSED. NOTE: ALL LS//100A KITS ARE INSTALLED/INCLUDED WITH A DIODE TO PREVENT REVERSED POLARITY AND THUS BURNOUT. SEE "OPTIONS" ON PAGE AP:03:06 FOR MORE DETAILS.

31 TAKE THE PRE-ASSEMBLED BOTTOM SECTION OF THE SECONDARY HULL AND TEST FIT TO DETERMINE IF THERE ARE ANY PROBLEMS IN FITTING. SOME FIBER OPTIC MAY HAVE TO BE TRIMMED, AND SOME WIRES MAY HAVE TO BE SHORTENED OR REPOSITIONED TO MAKE ROOM FOR THE PHONE JACK. THIS PROCEDURE IS DIFFICULT AND WILL REQUIRE SOME TIME. (SEE STEP 33).

32 GLUE THE FRONT END CAP AT THE FRONT OF THE SECONDARY HULL. BE SURE TO ALIGN THE CAP WITH THE THREE RECTANGULAR SECTIONS ON THE SIDES AND BOTTOM OF THE HULL.

33 TO PROPERLY DETERMINE THE LOCATION OF THE HOMING BEACON, FIRST TEST FIT THE BOTTOM SECTION WITH THE YELLOW LED MOUNTED, BUT NOT GLUED, IN THE SHUTTLE DOORS. THEN DRILL A 1/8" DIA. HOLE JUST ABOVE THE SHUTTLE DOORS. IF THE LED IS TOO SHORT, THEN ENLARGE THE HOLE IN THE SHUTTLE DOORS AND MOVE THE LED UP INTO THE NEW HOLE IN THE HULL.

34 IF ALL SYSTEMS ARE WORKING, AND NO ADJUSTMENTS ARE REQUIRED ON THE CIRCUIT BOARD, THE PRIMARY HULL TOP CAN BE FINALLY GLUED ON.

add impulse engines
29

30
phone jack
18A

32
putty & sand

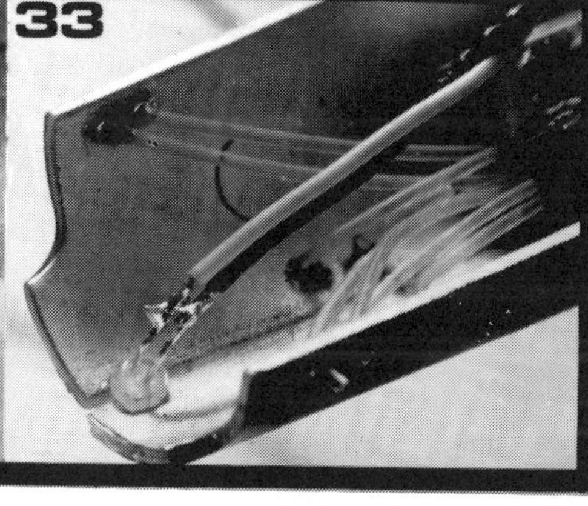
33

35 ONCE THE SHIP HAS BEEN FULLY ASSEMBLED, PUTTY THE FOLLOWING IMPORTANT LOCATIONS:

a. SECONDARY HULL CAP JOINTS
b. LOWER FRONT OF DORSAL AND TOP SEAM OF SECONDARY HULL
c. COMPLETE EDGE OF PRIMARY HULL (DO NOT GET PUTTY ON FIBER OPTIC)
d. DORSAL - PRIMARY HULL CONNECTION
e. PYLON - SECONDARY HULL CONNECTION
f. SEAMS ALONG SIDES OF SECONDARY HULL

USE (SQUADRON) GREEN PUTTY. IT IS MUCH MORE WORKABLE THAN TESTOR'S OR PACTRA PUTTY. ENAMEL NAIL POLISH WORKS WELL ALSO.

36 SAND THE ENTIRE SHIP, AND TRY TO AVOID THE INDIVIDUAL PROTRUSIONS OF FIBER OPTIC. IT MAY BE DIFFICULT, SO ALSO USE SMALL FILES AND STRIPS OF SANDPAPER. TRY PAINTING THE SHIP FLAT BLACK TO CHECK FOR ANY ROUGH OR UNEVEN SPOTS OVERLOOKED. SPEND SEVERAL HOURS WITH THIS PROCEDURE. TO OBTAIN A NEAR-PERFECT SURFACE.

37 ONCE THE SHIP HAS BEEN COMPLETELY SANDED, BEGIN TO MASK OFF ALL PLACES NECESSARY. WITH MASKING TAPE, TAPE THE NACELLE LEDs, AND THE SENSOR ARRAYS. WITH SMALLER PIECES OF TAPE, TAPE THE RUNNING LIGHTS AND HOMING BEACON. THE PORTS WILL BE PAINTED OVER AND THE FIBER OPTIC WILL BE TRIMMED FLUSH TO THE HULL'S SURFACE LATER. USING RUBBER CEMENT ON THE RUNNING LIGHTS, HOMING BEACON AND SENSOR ARRAYS IS POSSIBLE AS A MASK, AND CAN BE RUBBED OFF AFTER THE PAINT HAS DRIED.

38 REFER TO THE SUPERDETAILING SECTION FOR SHAPING AND ATTACHING NACELLE POWER SHIELDS.

39 REFER TO PAINTING SECTION FOR COLORS AND TECHNIQUES IN APPLYING PAINT.

40 YOU HAVE NOW COMPLETED THIS SECTION.

AFTER THE SHIP HAS BEEN COMPLETELY PAINTED AND WEATHERED, TRIM ALL EXTRUDING STRANDS OF FIBER OPTIC WITH FLUSH CUTTERS. MAKE SURE THE CUT IS CLEAN AND AS CLOSE TO THE SURFACE AS POSSIBLE. BE CAREFUL NOT TO SCRAPE ANY PAINT WITH THE CUTTERS.

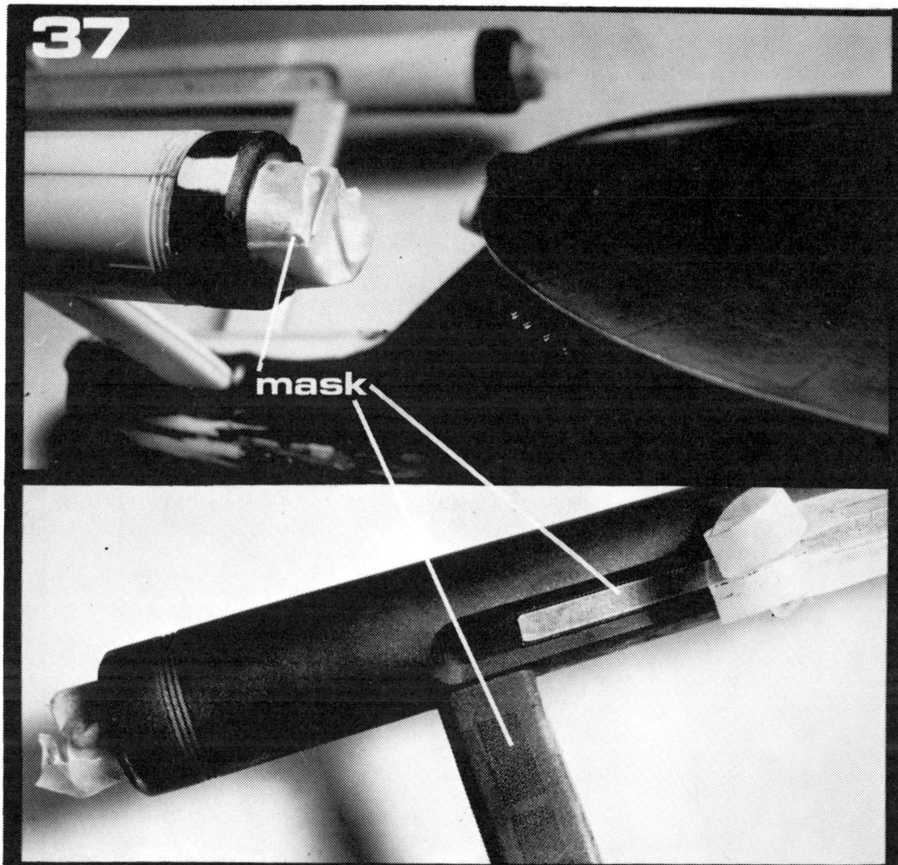
37
mask

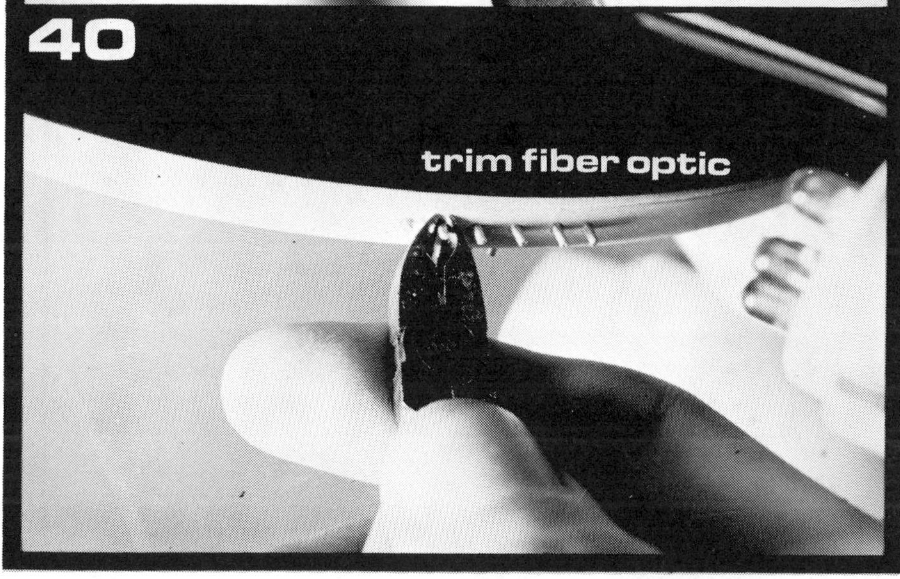
40
trim fiber optic

SUPPORT BASE/POWER CONNECTION

1 THE SUPPORT BASE DESIGNED FOR THE LS/100A STARSHIP SYSTEM CONSISTS OF THE FOLLOWING MATERIALS:*

1 - RADIO SHACK ADAPTER #274-325 (1/4" PHONE PLUG)
1 - RADIO SHACK PHONE JACK #372-252 (2 CONDUCTOR - OPEN CIRCUIT)
1 - RADIO SHACK MICROPHONE GOOSENECK #33-330
1 - RADIO SHACK PADDED DESK STAND #33-370
1 - TUBE SET OF QUICK-SET EPOXY

* YOU WILL ALSO NEED A SOLDERING IRON AND SOLDER.

2 BEGIN ASSEMBLY OF THE BASE BY TWISTING THE GOOSENECK ONTO THE PADDED DESK STAND, REPLACING THE 4" CHROMED TUBE WHICH COMES WITH THE STAND.

3 STRAND THE WIRES OF THE VOLTAGE ELIMINATOR (RADIO SHACK 270-1552, OR SEARS OR K-MART) RATED AT 6 VOLTS AND 250 mA OR 300 mA THROUGH THE GOOSENECK. FIRST, CUT OFF THE SMALL (NOT WALL PLUG) PLUG INTENDED FOR A TAPE PLAYER, ETC., AND STRIP OFF THE END OF THE WIRE.

4 REMOVE THE HOUSING SCREWED ONTO THE ADAPTER AND WITHOUT SOLDERING, ATTACH THE STRIPPED WIRES OF THE LEADS.

5 IF YOU HAVE NOT GLUED ON THE BOTTOM OF THE SECONDARY HULL ONTO THE SHIP, DETERMINE THE POLARITY OF THE PHONE JACK BY INSERTING THE PHONE PLUG (ADAPTER). IF THE PORTS LIGHT AND THE OTHER ELECTRONICS DO NOT, THE POLARITY IS REVERSED. CHANGE THE WIRES ON THE PHONE PLUG AND TRY AGAIN (NOTE: A DIODE HAS BEEN ADDED TO THE LS/100A BOARD TO PREVENT ANY REVERSE POLARITY FROM BURNING OUT THE ELECTRONICS). IF THE WIRES' ARRANGEMENT IS CORRECT, THEN PROCEED TO SOLDERING THE WIRES TO THE PHONE PLUG (ADAPTER). SEE STEP 8 BEFORE GLUING THE HULL INTO PLACE.

6 NOW, PLACE A CRUMPLED PIECE OF PAPER (OR THE LIKE) DOWN THE GOOSENECK FROM THE TOP ABOUT 2". THIS WILL BE USED TO PREVENT THE EPOXY GLUE FROM OOZING DOWN THE GOOSENECK WHEN APPLIED.

7 ONCE THE PHONE JACK HAS BEEN SOLDERED, WRAP A SMALL PIECE OF ELECTRICAL TAPE AROUND CONTACTS TO PREVENT SHORTING. POSITION INTO PLACE AS SHOWN, ADDING EPOXY INTO THE TOP OF THE GOOSENECK. WITH MASKING TAPE KEEP THE PLUG VERTICLE WHILE THE EPOXY IS DRYING. WHEN THE EPOXY IS ABOUT HALF DRY, TRY TO PLUG THE SHIP ONTO THE SUPPORT TO CHECK CLEARANCE. THE BOTTOM OF THE SHIP SHOULD BE FLUSH WITH THE TOP OF THE GOOSENECK, AND FULL ELECTRICAL CONNECTIONS SHOULD BE MADE, WITH NO FLICKERING OR OTHER PROBLEMS. IF SO, YOU MAY HAVE THE PHONE PLUG MOUNTED TOO DEEP INTO THE GOOSENECK.

8 THE TENSION OF THE PHONE PLUG IN THE PHONE JACK SHOULD BE ENOUGH TO KEEP THE SHIP FROM "SWIVELING" WHEN THE GOOSENECK IS BENT, BUT NOT SO MUCH THAT IT IS DIFFICULT TO REMOVE THE SHIP FROM THE BASE. THE TENSION CAN BE ADJUSTED BY FLATTENING OR EXPANDING THE LONG CONTACT ON THE PHONE JACK.

9 (OPTIONAL) PAINT THE GOOSENECK FLAT BLACK IF THE SHIP IS TO BE USED FOR FILMING. THE PHONE JACK/PHONE PLUG CAN ALSO BE MOUNTED AT ANY OTHER PLACE ON THE SHIP, DEPENDING ON YOUR PREFERENCE.

10 (OPTIONAL) TO HAND HOLD THE SHIP FULLY ILLUMINATED, OBTAIN A HEADPHONE EXTENSION CORD, ABOUT 25 FT. PLUG ONE END INTO THE SHIP AND THE OTHER END INTO THE TOP OF THE GOOSENECK. TURN OUT THE ROOM LIGHTS AND VIOLA!

11 (OPTIONAL) IF YOU CHOOSE TO USE DIFFERENT AND MORE POWERFUL LIGHTS FOR THE FIBER OPTIC LIGHT SOURCES, USE TWO POWER SUPPLIES AND A STEREO PHONE JACK AND PLUG CONNECTION (3 CONDUCTOR) AND CONNECT THE LIGHT SOURCE WIRES TO THE 2 OF THE 3 CONDUCTORS AS WELL AS THE LS/100A POWER WIRES TO THE OTHER 2 CONDUCTORS, WITH BOTH NEGATIVE LEADS SHARING THE CENTER CONDUCTOR. SEE "OPTIONS" IN THE LS/100A ASSEMBLY SECTION.

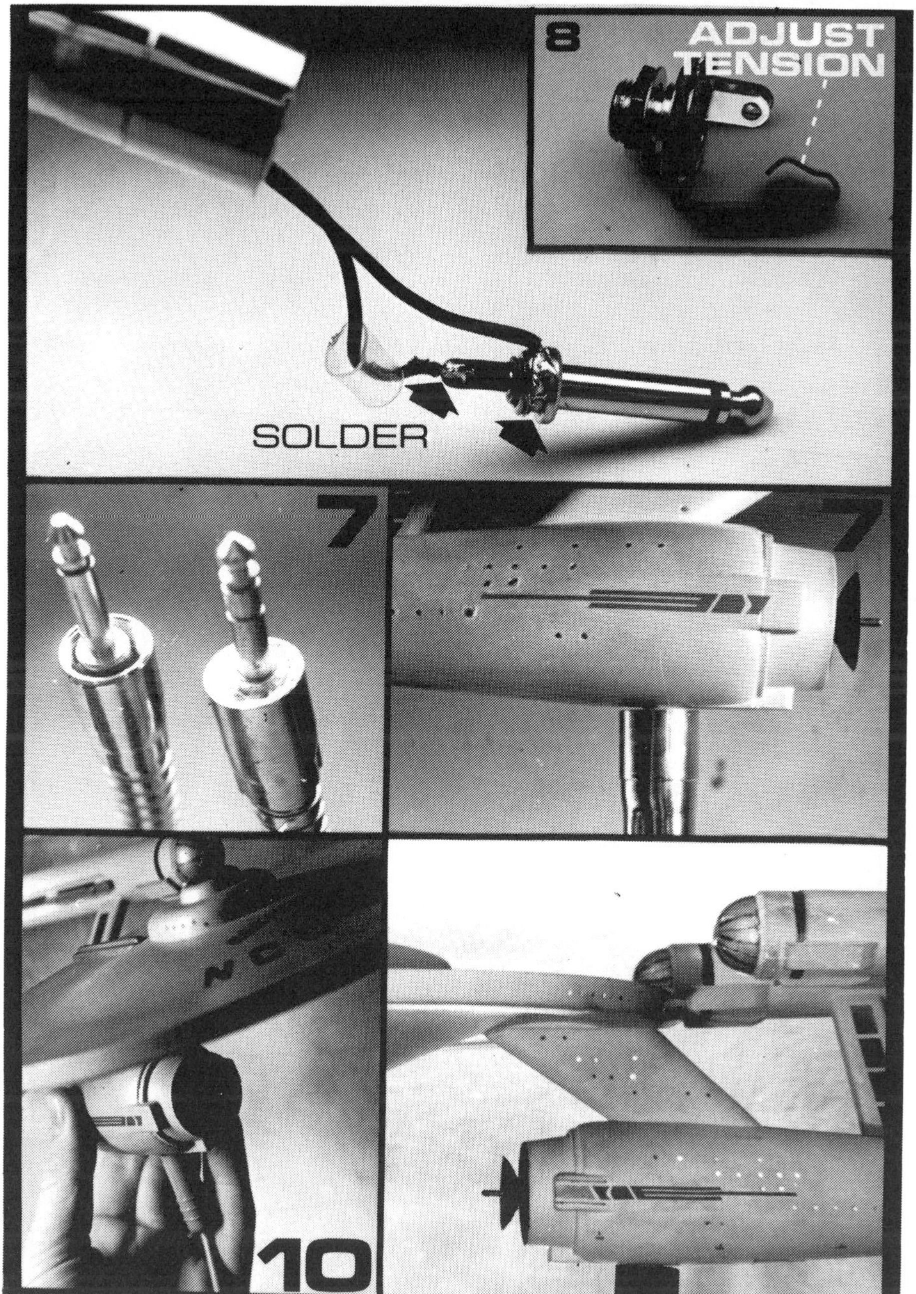
8
ADJUST TENSION
SOLDER
7
7
10

DECALS

1. TO PROPERLY DECAL YOUR HEAVY CRUISER STARSHIP, SUPPLEMENT THE AMT KIT DECALS WITH CHAMP DECALS (MODEL R.R.) WHICH ARE AVAILABLE AT ANY MODEL R.R. HOBBY STORE, AS LISTED BELOW:
   - -PRIMARY HULL, TOPSIDE (NCC-1701): A-19 BLOCK GOTHIC, BLACK, 1/2"
   - -OUTBOARD NACELLE SIDES NUMERALS: A-11 BLOCK GOTHIC, BLACK, 1/4"
   - -PRIMARY HULL, UNDERSIDE NUMERALS: X-98 O GAUGE, BLACK BUGBOARD NUMBERS

   WHEN USING BLOCK STYLE NUMERALS, THE #1 WILL REQUIRE A "HEAD" AT THE TOP: OTHERWISE IT WILL APPEAR AS THE LETTER "I". ALL OF THE OTHER ALPHABET AND NUMERAL CHARACTERS ARE ACCEPTABLE. THIS BLOCK STYLE IS ALSO ON ALL CURRENT U.S. NAVY SHIPS, AND THE CHOICE FOR THE ENTERPRISE SUGGESTS THE CONTINUANCE OF NAVY PRACTICES IN THE 23RD CENTURY.
2. OBTAIN SOME ESTES DECALS FOR THE FLYING ROCKET U.S.S. ENTERPRISE. USE ALL THE ALPHABET AND NUMERAL CHARACTERS INCLUDED WITH THE AMT STARFLEET PENNANTS. THESE DECALS ALSO INCLUDE THE PROPER RECTANGLES, FRAME NUMBERS, AND ARCHES FOR DETAILING. THE SET OF INSTRUCTIONS THAT COME WITH THIS KIT GRAPHICALLY SHOW PROPER PLACEMENT. THESE DECALS ALSO INCLUDE PORTS (BLACK AND WHITE) AND GRID LINES. HOWEVER, AS THE AMT KIT IS 1/600 SCALE, THE ESTES KIT IS ABOUT 1/650 SCALE, SO THE DECALS WILL BE SLIGHTLY SMALLER.
3. BELOW ARE LISTED THE HEAVY CRUISER STARSHIPS AND PROPER NUMERALS FOR EACH. THE STAR FLEET TECHNICAL MANUAL (SFTM) AND STAR TREK CONCORDANCE (STC) CONFLICT SOME, THEREFORE, BOTH ARE GIVEN:

DESIGNATION		SFTM		STC
USS CONSTELLATION*	(NCC)	1017	(NCC)	SAME
CONSTITUTION		1700		SAME
DEFIANT		NONE LISTED		1764
ENTERPRISE		1701		SAME
EXCALIBUR		1705		1664
EXETER		1706		1672
FARRAGUT		1702		1647
HOOD		1707		1703
INTREPED		1708		1631
KONGO		1710		NONE LISTED
LEXINGTON		1703		1709
POTEMKIN		1711		1702
REPUBLIC		1371		1373
VALIANT		1709		NONE LISTED
YORKTOWN		1704		1717

*THE U.S.S. CONSTELLATION WAS NUMBERED NCC 1017, AS IT APPEARED IN "DOOMSDAY MACHINE" IN THE FORM OF A BATTLE-DAMAGED AMT MODEL. THE "1017" WAS THE RE-ARRANGEMENT OF "1701" ENTERPRISE DECAL THAT CAME WITH THE KIT (SUCH A CHOICE OF NCC NUMBER HAS CAUSED MUCH CONFUSION). IF YOU CHOOSE TO USE THE LETTERING STYLE AS USED ON THE HEAVY CRUISERS IN THE STAR FLEET TECHNICAL MANUAL, REFER TO THE SPECIAL SECTION STARTING ON AP:03:43.

4. TO PROPERLY DUPLICATE THE SMALL FRAME NUMBERS ON THE SIDE OF THE SECONDARY HULL, AS EXPLAINED IN THE SUPERDETAILING SECTION, USE BLOCK GOTHIC STYLE NUMERALS MEASURING ABOUT 1/32" HIGH. THESE NUMERALS FROM AFT TO STERN: 102, 705, 1300, 1364, 1837. ALSO INCLUDE A FRAME MARK ABOVE EACH NUMBER.
5. ACQUIRE MICRO-SCALE DECALS RH-110, UP GENERAL STRIPING (RED) TO USE FOR ALL RED LINES AND ARCHES. WHEN APPLYING THE DECALS ALSO ADD SOME CHAMP DECAL-SET TO THE SURFACE WHERE THE DECAL IS TO BE APPLIED. THIS WILL SECURE THE DECALS AND ELIMINATE ANY AIR BUBBLES. WHEN DRY, SPRAY THE DECALS, AS WELL AS THE ENTIRE MODEL WITH SOME TESTORS DULLCOTE TO ASSURE PERMANENCE.

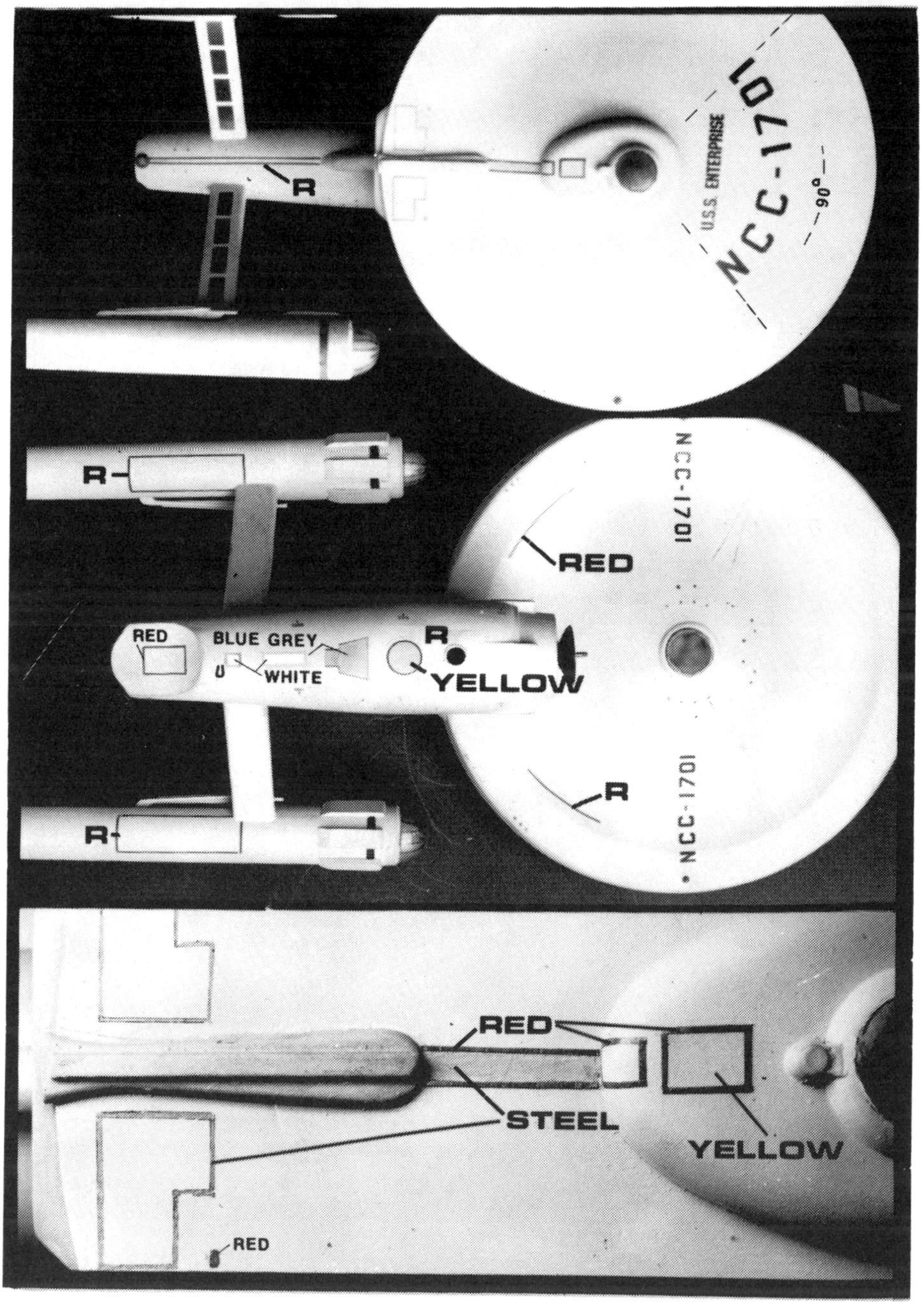
R
U.S.S. ENTERPRISE
NCC-1701
90°
R
NCC-1701
RED
RED
BLUE GREY
WHITE
R
YELLOW
R
NCC-1701
R
RED
STEEL
YELLOW
RED

SUPERDETAILING/ SURFACE DETAIL

THIS SECTION DEALS WITH THE SUPERDETAILING ON THE 11 FT. MODEL (143"), WHICH NOW HANGS IN THE "LIFE IN THE UNIVERSE" HALL AT THE SMITHSONIAN IN WASHINGTON, D.C. THESE SKETCHES ARE NOT DRAWN TO SCALE AND ARE INTENDED ONLY TO ILLUSTRATE VARIOUS DETAILS ON THIS MINIATURE.

1 THIS IS AN ISOMETRIC VIEW OF THE AFT STARBOARD SECTION OF THE SECONDARY HULL, SHOWING PORT PLACEMENT, ION POD, FRAME REFERENCE NUMBERS, HOMING BEACON AND SHUTTLECRAFT LANDING LIGHTS. ONE INTERESTING FEATURE ABOUT THE FRAME REFERENCE MARKS IS THAT THEY INDICATE THE LOCATIONS OF THE STRONGEST POINTS ON THE SHIP AND CAN BE USED FOR REFERENCE WHEN THE SHIP IS PLACED IN DRY DOCK (i.e., SHIP DRY DOCK SUPPORT CLAMPS).

2 THIS IS AN UNDERSIDE VIEW OF THE SECONDARY HULL SHOWING VARIOUS MARKINGS AS INDICATED. THE PURPOSE OF THESE PANELS IS SOMEWHAT UNCLEAR, HOWEVER IT IS LOGICAL TO ASSUME THAT THEY INDICATE WHERE THE THINNEST PARTS OF THE HULL ARE FOR RESCUE, OR REMOVABLE ACCESS PLATES FOR REPAIR OR SUPPLIES. THE EIGHT-SIDED PANEL MAY BE AN ACCESS TO THE TURBO-LIFT (BASED ON PAGE 177 OF THE MAKING OF STAR TREK). COLORS FOR THE PANELS ARE AS INDICATED.

IT IS ALSO IMPORTANT TO NOTE THE LIPPED EDGE AS INDICATED BY THE LETTER "L".

3 THIS FIGURE ILLUSTRATES THE STARBOARD VIEW, SHOWING PORT PLACEMENT, LOCATION OF FRAME REFERENCE NUMBERS, PROPER ANTENNA (MAIN SENSOR), DIMENSIONS, SHIP FUNCTION LIGHT, HOMING BEACON AND ION POD. ALSO NOTE THE SHARPER ANGLE OF THE LEADING EDGE OF THE DORSAL. THIS ANGLE IS DIFFERENT THAN THAT OF THE HEAVY CRUISER BLUEPRINTS AND THE AMT KIT.

THE TIP OF THE MAIN SENSOR IS NARROWER AS INDICATED BY THE ARROW.

NOTE THE DETAIL AHEAD OF THE STAR FLEET PENNANT. THE CONCENTRATED LINES ARE ENGRAVED INTO THE SURFACE, WHILE THE TOP AND BOTTOM LINES ARE RELIEFED. (THIS SKETCH IS AN ARTIST'S RENDERING, AND INTENDED FOR SCALE MODELING PURPOSES ONLY).

StarFleet Division

ADVANCED ASSEMBLY PROCEDURE (A)APHC-3

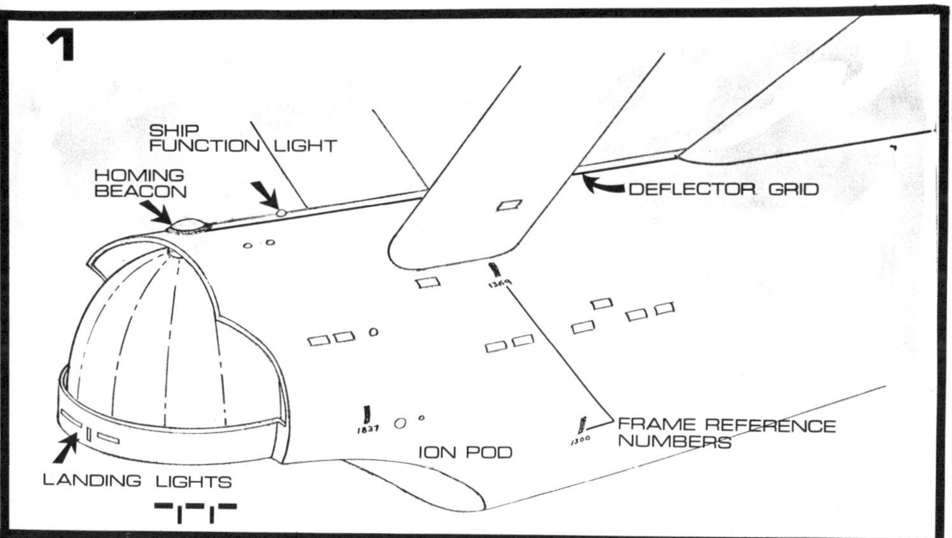

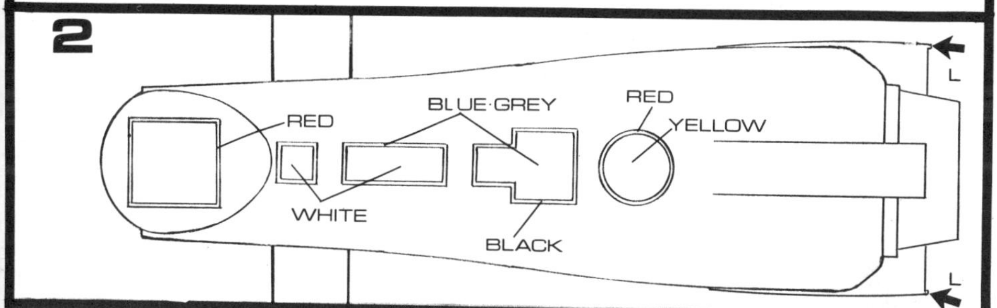

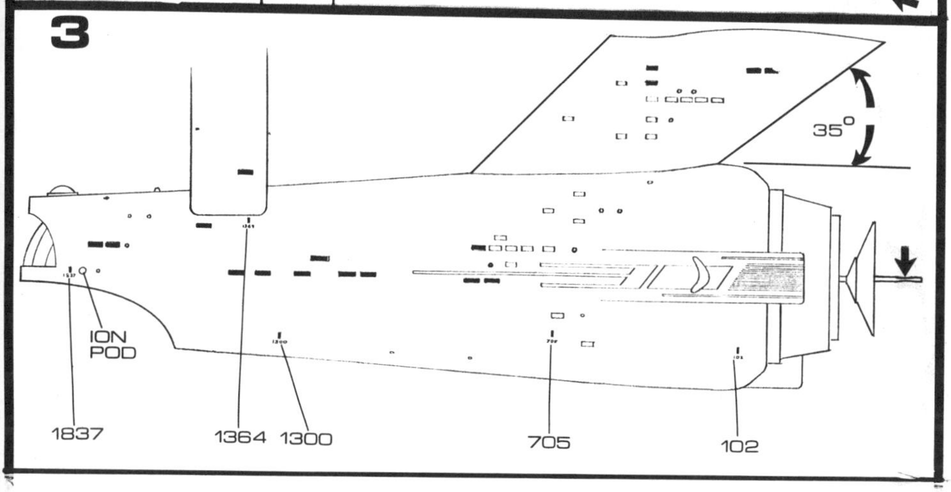

4 THIS FIGURE ILLUSTRATES THE PROPER SHAPE OF THE POD REACTOR. PLEASE NOTE THE LOCATION OF TWO RIVETS SECURING THE REACTOR. TO DUPLICATE THESE RIVETS, USE HO SCALE MODEL R.R. TRACK SPIKES OR TINY ROUND-HEADED TACKS. ALSO NOTE THAT THERE ARE COOLING FINS ON THE PIPE SECTION, HOWEVER THESE FINS ARE MISSING ON THE 4 FT. STUDIO MINIATURE.

5 THIS SKETCH ILLUSTRATES THE FORWARD SECTION OF THE NACELLE. THE SPIKE ONLY APPEARED ON THE EARLY VERSION OF THE SHIP (i.e., "WHERE NO MAN HAS GONE BEFORE" AND "THE MENAGERIE").

NOTE THE RESTRAINING CLIP AS INDICATED BY THE ARROW. THIS APPEARS ON THE AMT KIT ACCURATELY. DUE TO DAMAGE ON THE SMITHSONIAN MINIATURE OVER THE YEARS, WHEN THE NACELLE DOMES WERE REPLACED THE RESTRAINING CAPS WERE OVERLOOKED DURING RESTORATION.

THE NACELLE (POWER) SHIELDS ARE SHOWN HERE IN THEIR ACCURATE FORM. TO DUPLICATE THIS ON THE MODEL, CUT 1" STRIPS OF PLASTRUCT TB-10, ROUNDING OFF THE LEADING EDGE AND ANGLING THE TRAILING EDGE AT ROUGHLY 30°.

6 THIS DIAGRAM ILLUSTRATES THE PROPER SHAPE OF THE UPPER FOUR DECKS AND BRIDGE. IT ALSO SHOWS THE PLACEMENT OF PORTS AND TURBO-LIFT. THE RECTANGLE ON THE LOWER SKETCH HAS BEEN SUGGESTED AS EITHER A PERMEABLE WINDOW OR SLIDING DOORS FOR THE PHYSICS TELESCOPE/SEXTANT, AS IN THE HEAVY CRUISER BLUEPRINTS, SHEET NO. 5.

THE DOTTED LINE ILLUSTRATES THE LARGER BRIDGE DIMENSIONS ON THE 1966 DESIGN. ALSO NOTE THE INCREASED CURVATURE OF THE TOP SENSOR ARRAY.

4

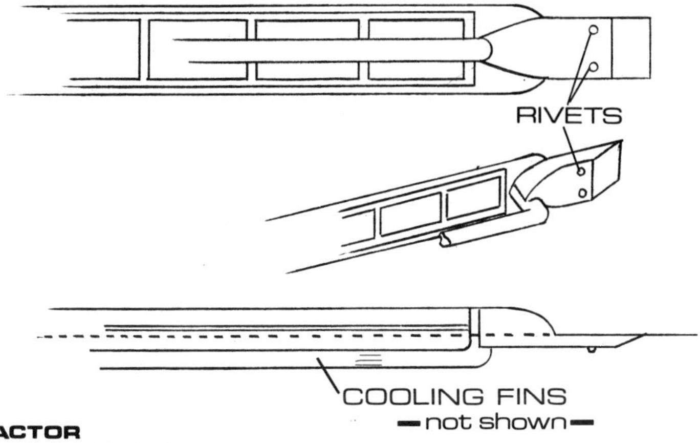

POD REACTOR

5

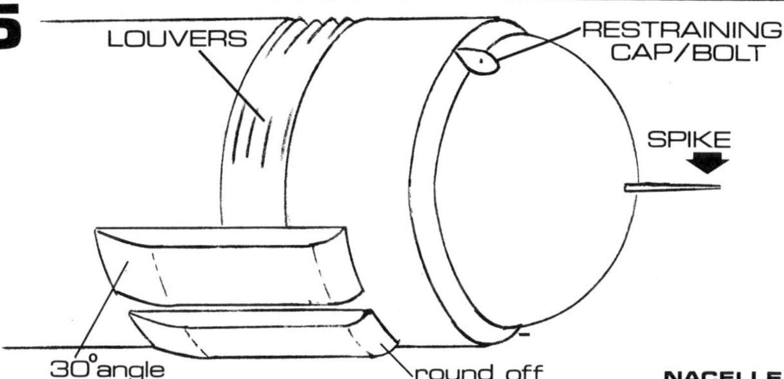

NACELLE

6

BRIDGE

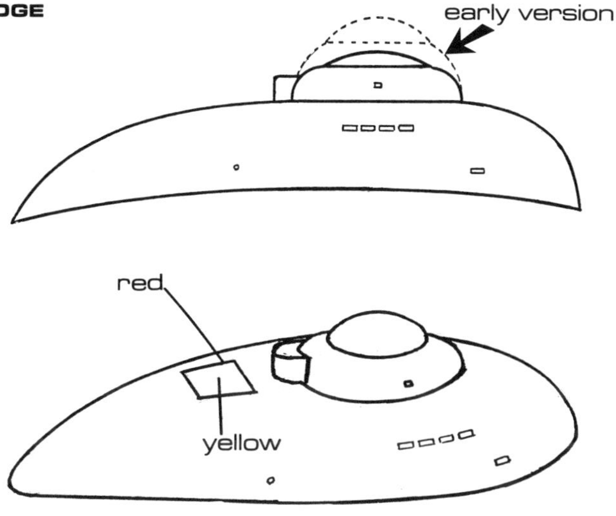

OPTIONS / STUDIO MINIATURES

BELOW ARE LISTED SOME GUIDELINES TO PROTOTYPE MODELING OF THE 1966 11 FT. MINIATURE AND 4 FT. MINIATURE. THESE DETAILS ARE CONDENSED AND SERIOUS MODELERS WILL HAVE TO CONSULT EARLY FILM CLIPS OR OBTAIN A COPY OF TREK #5 (NOW OUT OF PRINT). ALSO CHECK THE PHOTO NOVELS/PAPERBACKS OF THE EPISODES LISTED. ALSO CHECK #13 ON PAGE AP:03:50.

1. TO DUPLICATE THE FIRST UNREVISED MINIATURE (1966) USED IN THE EPISODE "WHERE NO MAN HAS GONE BEFORE", AND "THE CAGE" ("MENAGERIE"), NOTE THE FOLLOWING DETAILS:
   - A. THE REAR NACELLE CAPS HAD NO DOME: INSTEAD, 6 ROWS OF DOTS ON THE CAP'S CONTOUR.
   - B. THE BRIGHT RED NACELLE DOMES HAD TWO GOLD SPIKES MOUNTED IN THE CENTER OF EACH DOME. TO DUPLICATE THIS, USE A "FINIAL" FOR MODEL R.R. BUILDINGS (SCALE STRUCTURES #650-2288) OR SIMILAR.
   - C. THE HULL WAS A LIGHTER COLOR; WHITE WITH A SILVERY COATING.
   - D. THERE WERE NO SCREENS ON THE INBOARD SIDE OF THE SUPPORT PYLONS.
   - E. THE INTERCOOLERS HAD NO SCREENS.
   - F. POD REACTORS WERE SILVER, AND HAD LITTLE DETAIL. THE OTHER SCREENS, VENTS AND NACELLE END CAPS ARE A DARKER SHADE OF THE HULL COLOR. THE IMPULSE ENGINES WERE THE HULL COLOR AND NOT BLUE-GREY.
   - G. THE UNDERSIDE RUNNING LIGHTS WERE FARTHER BACK AND WHITE. THE HOMING BEACON FLASHED AT THE SAME RATE AS RUNNING LIGHTS.
   - H. THE MAIN SENSOR WAS A MUCH LARGER DIAMETER AND A BRIGHTER RED-COPPER. THE DISH WAS MORE BOWL-LIKE.
   - I. THE BRIDGE SECTION WAS LARGER, AND TALLER, WITH A BRIGHT (VIEWSCREEN) RECTANGLE MOUNTED ON THE FRONT (EXTERIOR). USE ANOTHER GUMBALL CAPSULE TO DUPLICATE THIS TALLER BRIDGE, AND USE A LONGER PLEXIGLAS ROD.
   - J. THE PRIMARY HULL TOPSIDE HAD TWO CROSSHATCHED PANELS THAT WERE MOUNTED AROUND THE TOPSIDE RUNNING LIGHTS (SEE THE PREVIOUS SECTION).
   - K. NO DETAIL ON THE SHUTTLECRAFT BAY DOORS. NO LINES BETWEEN DOOR SECTIONS.
   - L. THIS 11 FT. MINIATURE ALSO HAD PORT LIGHTS ONLY ON THE STARBOARD SIDE, BECAUSE OF A LACK OF MONEY FOR THE SERIES, AND THE DIFFICULTY (AT THAT TIME) TO CONTAIN THE POWER WIRES. TO FILM THE "OTHER SIDE", LETTERING ON THE STARBOARD NACELLE WAS FLIPPED OVER TO READ BACKWARDS, AND THEN THE FILM WAS FLIPPED OVER TO SIMULATE THE PORT SIDE. YOU MIGHT ALSO NOTE THE 2 MISSING PORT SIDE RECTANGLES (OF 4) ON THE TOPSIDE OF THE PRIMARY HULL.
2. HERE ARE SOME DETAILS UNIQUE TO THE 4 FT. MINIATURE.
   - A. THE PORTS WERE PAINTED LIGHT YELLOW OR BLACK; NO INTERIOR LIGHTS. PORTS ON BOTH SIDES.
   - B. THERE WERE A SET OF RUNNING LIGHTS (AS ON YOUR SHIP) PLUS TWO SMALL WHITE FORMATION LIGHTS 2-3".
   - C. THE REAR NACELLE "BALLS" ARE MUCH LARGER THAN THOSE LATER ADDED TO THE 11 FT. MINIATURE.
   - D. THE HOMING BEACON, WHEN OFF, RESEMBLED AN UNLIT CAR HEADLIGHT, AND ONE FROM A 1/24 SCALE CAR WOULD WORK AS WELL. WITH 1/32" GRAPHIC TAPE MAKE A MOUNTING RING.
   - E. MANY MORE MARKINGS WERE ADDED TO THIS SMALLER MINIATURE AND ARE TOO NUMEROUS TO MENTION. PLEASE CHECK THE POSTER LISTED IN REFERENCES FOR DETAILS.

SUPERDETAILING/ SFTM: 379260

1. TO FAITHFULLY REPRODUCE THE CLASS I HEAVY CRUISER ACCORDING TO THE SPECIFICATIONS OF STAR FLEET COMMAND, FIRST REFER TO THE FOLLOWING REFERENCES:
   BOOKLET OF GENERAL PLANS, U.S.S. CONSTITUTION CLASS - ALL 12 SHTS
   STAR FLEET TECHNICAL MANUAL TM: 379260
   REFER TO PAGES:
   TO:00:10:00 and 00:10:10
   01:04:00 and 01:04:10 and 01:04:11
   01:06:11
   03:08:10
   03:10:20
   03:11:20
   03:17:00 and 03:17:01
2. THE FOLLOWING INFORMATION OUTLINES THE DETAILS ON THESE SHIPS ACCORDING TO STAR FLEET COMMAND, AS COMPARED TO OTHER SOURCES:

PRIMARY HULL

A. DEFLECTOR SHIELD GRID: BLUEPRINTS GIVE 24 RADIAL LINES AND 5 CIRCLES. THE AMT KIT HAS 28 RADIAL LINES AND 6 CIRCLES. THE AMT KIT ALSO DOES NOT HAVE ANY LINES ON THE HULL UNDERSIDE. LINES ARE ETCHED INTO SURFACE.
B. BOW LIGHT: TOP SIDE OF HULL. MOUNTING RING AT BASE.
C. RUNNING LIGHTS: (4) SAME LOCATION AND COLOR ORDER AS STANDARD NAVIGATION LIGHTS - 20TH CENTURY AIRCRAFT. LIGHTS HAVE SMALL MOUNTING RING AT BASE.
D. FORMATION LIGHTS: (2) NO MOUNTING RINGS. COLORS CLASSIFIED BY SFC (SMALL FORMATION LIGHTS APPEAR ON THE 4 FT. STUDIO MINIATURE AND ON THE 1966 VERSION OF THE 11 FT. MINIATURE, ON THE UNDERSIDE ONLY).
E. WINDOWS/PORTS: CHECK PLACEMENT ON BLUEPRINTS IN COMPARISON TO STAR TREK FILM CLIPS.
F. VARIATION IN UPPER FORE DECK'S HULL SECTION DIMENSION.
G. BLUEPRINTS SHOW GREATER ANGLE TO THE EDGE OF HULL.
H. PHASER BANKS: LOCATED ON PORT AND STARBOARD SIDES OF TOP SIDE OF HULL: FORWARD PHASER BANKS LOCATED UNDERSIDE OF HULL, FORWARD OF SENSOR ARRAY.
I. THRUST PORTS/IMPULSE ENGINES: FOUR SETS OF DOUBLE THRUST PORTS (INSTEAD OF TWO AS ON STUDIO MINIATURES).
J. TURBO LIFT/BRIDGE (EXTERIOR): FLUSH WITH BRIDGE EXTERIOR SURFACE.
K. SPACE AND PLANETARY SENSOR ARRAY: SAME AS TOP SENSOR ARRAY (THE 11 FT. MINIATURE HAD A SMALL SILVER TIP EXACTLY AT THE CENTER).
L. SEARCH SENSORS AND DEFLECTORS: LOCATED AT EXACT FRONT OF HULL (SIDE). USE "MV LENSE #516-119, 4 MM DIAMETER (USE 3) AVAILABLE AT MOST MODEL RAILROAD HOBBY STORES. ILLUMINATE FROM BEHIND WITH GRAIN-OF-WHEAT BULBS (THE CENTER SENSOR SHOULD BLINK, AS SHOWN AT THE INTRO SEQUENCE OF EVERY STAR TREK EPISODE).

SECONDARY HULL

A. PORTS: CHECK BLUEPRINTS AGAINST FILM CLIPS.
B. TRACTOR BEAM: SCRATCH BUILD USING PLASTIC TUBING AND PLASTIC SPHERE 3/16" DIAMETER.

NACELLES/MAIN PROPULSION

A. SPACE ENERGY/MATTER SOURCE: MADE FROM NACELLE BALLS FROM ROMULAN BOP KIT. TRIM TO FIT.

B. OUTLET FLOW SENSOR: HAS NO COOLING FINS AS ON KIT POD REACTOR. NO BOLT HEADS ON MOUNTING PLATE.

C. FINAL STAGE INTERCOOLERS P/S: TABS AT BASE; NO SCREENS AS ON KIT (THE 4 FT. MINIATURE HAD NO SCREENS EITHER, HOWEVER, THE 11 FT. MINIATURE DID, ALTHOUGH NOT NOTICE-ABLE BECAUSE SCREEN WAS PAINTED BLACK AND INTERCOOLER WAS MOUNTED AT AN OBLIQUE ANGLE).

LETTERING AND NUMERALS

A. NAVAL CONSTRUCTION CONTRACT NUMERALS ON TOPSIDE OF PRIMARY HULL, EXTENDED BLOCK STYLE, SIMILAR TO U.S. NAVY, 20TH AND 21ST CENTURY. MICROGRAMMA EXTENDED, 36 PT (11.7 MM), ALSO ACCEPTABLE TO STAR FLEET SPECIFICATIONS.

B. SHIP DESIGNATION, TOPSIDE OF PRIMARY HULL, NCC NUMERALS, UNDERSIDE OF PRIMARY HULL, MICROGRAMMA EXTENDED, 12 PT (3 MM).

C. SHIP DESIGNATION, UNDERSIDE OF SECONDARY HULL, MICROGRAMMA EXTENDED, 8 PT (1.6 MM).

D. IDENTIFICATION NUMERALS, NACELLE OUTBOARD SIDES CENTERED IN STAR FLEET IDENTIFICATION PENNANT, P/S, MICROGRAMMA EX-TENDED, 16 PT (4.4 MM), EUROSTYLE EXTENDED CAN BE SUBSTI-TUTED.

E. MICROGRAMMA EXTENDED DRY TRANSFER LETTERING AVAILABLE IN SHEETS OF VARIOUS POINT SIZES LISTED AT ANY GRAPHIC OR ART STORE. ASK FOR ZIPATONE: 632-36CN, 632-12CN, 632-8CN AND NORMATYPE 61 6038-16CL (EUROSTYLE EXTENDED).

MARKING AND PAINTING SPECIFICATIONS/CLASS I STARSHIPS

A. CORRECTIONS

ACCORDING TO UNITED FEDERATION REPRESENTATIVE, FRANZ JOSEPH, THE FOLLOWING COLORS WERE NOT PROPERLY REPRODUCED BY BALLANTINE BOOKS IN PRODUCING THE TERRAN CADET VERSION OF THE SFTM: 379260 PLEASE REFER TO PAGE TO:00:10:10 TO NOTE THE FOLLOWING:

SF08UC SAND - EXCESSIVE PINK
SF20UC MIDNIGHT - NOT DARK ENOUGH
SF29UC PLATINUM - SOFT GREY INSTEAD OF ALUMINUM

B. OFFICIAL SPECIFICATION RELEASE

THE FOLLOWING PAGE WAS RECEIVED BY PAUL M. NEWITT, CHIEF OF FLEET OPERATIONS, STAR FLEET DIVISION (20TH CENTURY), FROM FRANZ JOSEPH, UNITED FEDERATION REPRESENTATIVE (23RD CENTURY), STAR FLEET COMMAND, UNITED FEDERATION OF PLANETS ON STARDATE 7808.18. IT IS WITH GREAT ENTHUSIASM THAT STAR FLEET DIVISION IS ABLE TO PRESENT THIS HIGHLY OFFI-CIAL SPECIFICATION DATA.

REC'D-SFD/tSD 7808.18 :

OFFICIAL USE ONLY
O MARK O R5PC

STAR FLEET COMMAND
STAR FLEET HEADQUARTERS
UNITED FEDERATION OF PLANETS

FRANZ JOSEPH
UNITED FEDERATION REPRESENTATIVE

STARDATE 7306.29

PAUL M. NEWITT
STAR FLEET DIVISION
UNITED NATIONS/EARTH/SOL SYSTEM

1. IN REPLY TO YOUR INQUIRY AS TO THE MARKING AND PAINTING SPECIFICATION FOR CLASS I STARSHIPS, WE CAN PROVIDE THE FOLLOWING GUIDE, TRUSTING YOU REALIZE SOME COLORS ARE NATURAL - NOT 'PAINTED:'

NO. SF01UC SILVER	CONTROL REACTOR LOOP; FINAL STAGE INTER-COOLERS; FWD. RING EDGES OF DEFLECTOR ENERGY PUMP AND DEFLECTOR PULSE AMPLIFIER (MK-IX); HANGAR DOORS (MK-IX); POWER STAGE MAGNATOMIC FLUX CHILLERS; MAIN ENERGY STAGE MAGNATOMIC FLUX CHILLERS; PHASER UNITS; PHOTON TORPEDO EJECTOR NOZZLES
NO. SF05UC COPPER	MAIN SENSOR UNIT
NO. SF06UC WHITE	SECONDARY HULL FIRST BODY RING FWD. OF INTERCONNECTING DORSAL
NO. SF10UC RED	BASIC STAR FLEET IDENTIFICATION PENNANT
NO. SF11UC ORANGE	SPACE ENERGY/MATTER SINK (OPERATING); LARGE CIRCULAR PLATE ON FWD. BOTTOM OF SECONDARY HULL (MK-IX)
NO. SF20UC MIDNIGHT	MAGNATOMIC FLUX CONSTRICTION FIRST STAGE RINGS
NO. SF22UC CORN	WING EMBLEM OF STAR FLEET IDENTIFICATION PENNANT; AND WHEN OPERATING - SPACE ENGERY/MATTER SOÙRCE, PRIMARY HULL UPPER & LOWER SENSOR PLATFORM SHIELDS; SECONDARY HULL HOMING BEACON SHIELD (MK-IX); ALL EXTERNAL LIGHTS; ENVIRONMENTAL SYSTEM REACTORS
NO. SF29UC PLATINUM	BASIC EXTERNAL BODY (GROUND) COLOR; AND WHEN NOT OPERATING - SPACE ENERGY/MATTER SINK, ALL EXTERNAL LIGHTS, PRIMARY HULL UPPER & LOWER SENSOR PLATFORM SHIELDS, SECONDARY HULL HOMING BEACON SHIELD (MK-IX), ENVIRONMENTAL SYSTEM REACTORS
NO. SF32UC BLACK	ALL NUMERALS AND LETTERING; INTERIOR SURFACES OF DEFLECTOR ENERGY PUMP AND DEFLECTOR PULSE AMPLIFIER (MK-IX), INTERIOR SURFACE OF SPACE ENERGY/MATTER MATTRIX RESTORATION COWL
NO. SF33UC MED DARK GRAY	MAGNATOMIC FLUX CONSTRICTORS; CONCAVE SURFACE BEHIND THE MAGNATOMIC FLUX CHILLERS; FINAL STAGE MAGNATOMIC FLUX CHILLERS; SPACE ENERGY/MATTER MATTRIX RESTORATION COWL; PRIMARY HULL UPPER REAR RAISED VENTRAL FIN; IMPULSE POWER UNIT BODY; ANTI-MATTER CHILLERS ON SIDE OF PROPULSION UNIT SUPPORT PYLONS

NOTE: WHEN ENERGIZED, THE FORMATION LIGHTS SHOW NO. SF10UC RED, NO. SF12UC YELLOW, NO. SF14UC OLIVE, AND NO. SFL6UC BLUE. THE ORDER OF LIGHTS IS CLASSIFIED AND CONTAINED IN THE STAR FLEET COMMAND MANUAL. YOUR CIVILIANS CALL THE ENVIRONMENTAL SYSTEM REACTORS 'WINDOWS' AND 'PORTS'

PAINTING & COLORS

PAINTING YOUR MODIFIED HEAVY CRUISER STARSHIP IS NO EASY TASK, AND IS EQUALLY AS ADVANCED. TO ACHIEVE A PROFESSIONAL FINISH IT IS HIGHLY SUGGESTED FOR YOU TO OBTAIN A FAIRLY HIGH QUALITY AIRBRUSH. THE ONE PICTURED HERE IS A BINKS WREN-B, AN AIRBRUSH OF ABOUT $30-$40, AND WHEN USED WITH A COMPRESSOR, SUPPLYING AIR PRESSURE OF 30-50 POUNDS, NEAR PERFECT RESULTS ARE OBTAINABLE.

BELOW ARE THE COLORS AND FORMULAS SUGGESTED FOR PAINTING THE 11 FT. USS ENTERPRISE MINIATURE AS IT APPEARED IN 1968.

F=FLOQUIL (LACQUER) PAINT H=HUMBROL (ENAMEL) PAINT

PART	COLOR	RECOMMENDED PAINT	FORMULA
HULL POD REACTOR, INTERCOOLERS	SOFT GREY	F: 9 PRIMER 11 REEFER WHITE 144 PLATINUM MIST H:64 LIGHT GREY 34 WHITE 56 ALUMINUM	4 PARTS 4 PARTS 1 PART 4 PARTS 4 PARTS 1 PART
SCREENS NACELLE LOUVERS INTERCOOLER SCREENS	WEATHERED BLACK	F:17 WEATHERED BLACK H:92 IRON GREY	1:1 THINNER 1:1 THINNER
VENTS/ REAR NACELLE CAP	WEATHERED SILVER	F:100 OLD SILVER 9 PRIMER H: 87 STEEL GREY	1:1
REAR NACELLE CAP/CONTOUR	BLACK	F: 10 ENGINE BLACK H: 33 BLACK	NONE NONE
NACELLE "BALL"	WHITE	F: 11 REEFER WHITE H: 34 WHITE	NONE NONE
IMPULSE ENGINES DORSAL TOP PLATE	BLUE-GREY	F:130 S.P. LET. GREY H: 87 STEEL GREY 53 GUN METAL	NONE 5:1 GUN METAL
THRUST PORTS	CARBON BLACK	F:119 GRAPHITE H: 85 COAL BLACK	NONE NONE
ANTENNA (MAIN SENSOR) DISH-FRONT	DULL COPPER	F:105 COPPER 17 WEATHERED BLACK H: 55 FIRE BRONZE 53 GUN METAL	10:1 BLACK 8:1 GUN METAL
FRONT HULL CAP BEHIND MAIN SENSOR	†	F:108 GUN METAL H: 67 TANK GREY	NONE NONE
PORTS: LIGHTED	LT YELLOW*	F: 31 REEFER YELLOW 85 ANTIQUE WHITE	 8:1 WHITE
PORTS: UNLIGHTED	DK GREY*	F: 12 REEFER GREY 10 ENGINE BLACK H: 67 TANK GREY	 1:1 BLACK NONE
NACELLE DOMES EXT.	††RED TO AMBER-SEE SCHEMATIC	F:135 DAYLIGHT RED H: 60 SCARLET OR 51 SUNSET RED	SEE NOTES
NACELLE DOMES INT.	††YELLOW GOLD METALLIC	F:104 BRASS) 1:1 31 YELLOW) H: 54 BRASS 24 TRAINER YELLOW	SEE NOTES

NOTES:

- IMPORTANT! DO NOT MIX FLOQUIL WITH HUMBROL PAINT. BOTH BRANDS HAVE BEEN LISTED FOR EACH COLOR (CHOOSE ONLY ONE).
- BE SURE TO OBTAIN A SUBTLE APPEARANCE OF ALL THE VARIOUS COLORS. THIS WILL CREATE GREATER REALISM AND BETTER TONAL VARIATION.
- THE HULL COLOR IS A GREY VALUE BETWEEN BATTLESHIP GREY AND WHITE PAINTED OVER A HULL MADE OF "TRITITANIUM", WHICH GIVES THE SHIP A SILVERY APPEARANCE SOMETIMES.
- SPACE DEBRIS, DUST AND OTHER MATTER CLINGS TO THE ENGINES, ESPECIALLY NEAR THE MATTER ACQUISITION FANS, DESPITE THE DEFLECTOR GRID. TO SIMULATE THIS, GENTLY RUB GRAPHITE ON THE ENGINES. A GENERAL OVERSPRAY OF "FLOQUIL GRIME" THINNED 1:15 WITH THINNER, WORKS WELL ALSO. HOWEVER, PREVENT ANY WEATHERING FROM GETTING ON PORTS.
- SOME RETOUCHING HAS BEEN DONE ON THE 11 FT. MINIATURE AFTER IT ARRIVED AT THE SMITHSONIAN (WASHINGTON, D.C.). THE FOLLOWING IS INCORRECT: NACELLE DOMES TOO RED: MAIN SENSOR AND RINGS FOLLOWING TOO BRIGHT: THE REAR NACELLE CAP CONTOUR IS DARK GREY AND NOT BLACK.
- THIN ALL PAINT SOMEWHAT IF USING AN AIRBRUSH. PAINTING TECHNIQUES ARE UP TO YOU AND MAY REQUIRE SOME PRACTICE. MASKING IS ESSENTIAL WHEN AIRBRUSHING.
- IF YOU WISH TO ADD ANY BATTLE DAMAGE, SPRAY THE AREA DAMAGED FLAT BLACK, THEN FOG OVER WITH THE HULL COLOR. THIS GIVES A SUREALISTIC 3-D LOOK.
- SEE SUPERDETAILING SECTION ON PAGE AP:03:36 FOR DECALING AND EXTERIOR DETAILS SUCH AS BOTTOM PANELS - SECONDARY HULL.
- † THIS PAINT IS A COPPER COLOR SHADED WITH BLACK GRIME.
- †† TO PAINT THE INTERIOR OF THE NACELLE DOMES, START WITH DRAWING 12-14 FAN BLADES WITH A FINE TIP FELT PEN, THEN COAT OVER WITH AMBER METALLIC PAINT (NAIL POLISH). TO PAINT THE EXTERIOR OF THE NACELLE DOMES, LIGHTLY STREAK (DRY BRUSH) FLAT RED PAINT ON THE SURFACE TO GIVE THE ILLUSION OF SWIRLING ENERGY FROM THE AMBER METALLIC PAINT SHOWING THROUGH THE DOME. CHECK FILM CLIPS FOR DETAILS.
- * IF NOT USING FIBER OPTIC METHOD.

SUBSTITUTES:

NOT ALL HUMBROL COLORS ARE AVAILABLE AT ALL HOBBY STORES. HERE ARE SOME SUBSTITUTES FROM THE HUMBROL MILITARY COLORS GROUP: 34 WHITE = HB 11 UNDERSIDE WHITE: 64 LIGHT GREY = HU 6 LT. GREY; 67 TANK GREY = HM 4 GERM. PANZ. GREY; 74 LINEN = HB 16 CLEAR LINEN DOPED; 79 BLUE GREY = HV 3 NEUTRAL GRAY; 85 COAL BLACK = HB 10 NIGHT BLACK; 11 SILVER FOX = HB 14 AIRFRAME SILVER; and 53 GUN METAL = H5 217 STEEL. PLEASE ASK FOR A HUMBROL 80 COLOUR CARD WHERE HUMBROL PAINTS ARE SOLD. IT INCLUDES 80 COLOR SAMPLES. QUITE HELPFUL. A FLOQUIL COLOR CARD IS INCLUDED IN "PAINTING MINIATURES", A BOOK BY FLOQUIL AVAILABLE AT MOST HOBBY STORES.

BINKS' WREN-B AIRBRUSH

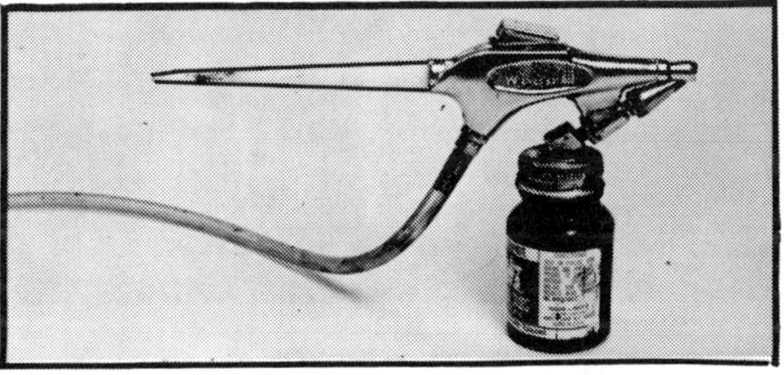

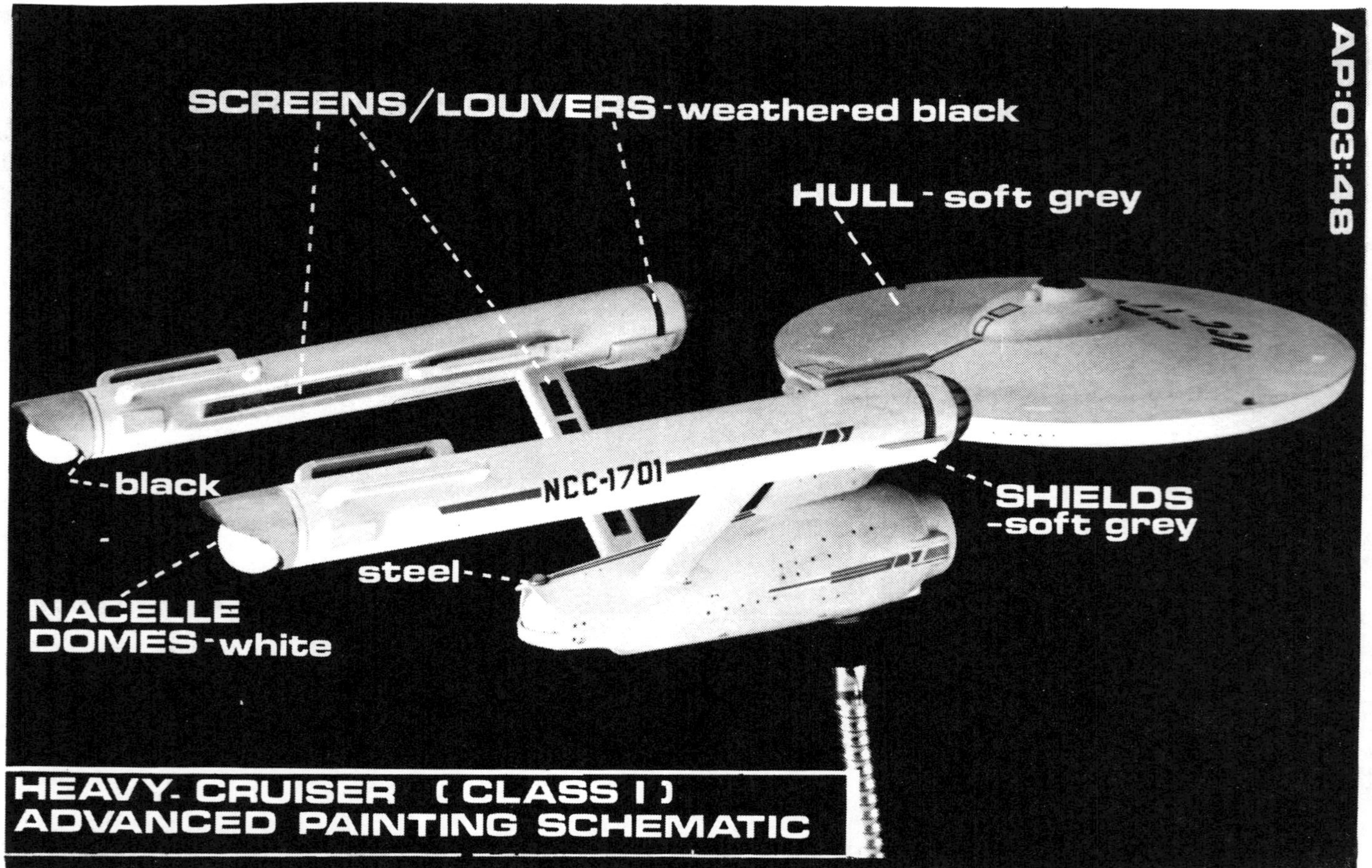
AP:03:48
SCREENS/LOUVERS-weathered black
HULL-soft grey
NCC-1701
black
SHIELDS
-soft grey
steel
NACELLE
DOMES-white
HEAVY-CRUISER (CLASS I)
ADVANCED PAINTING SCHEMATIC

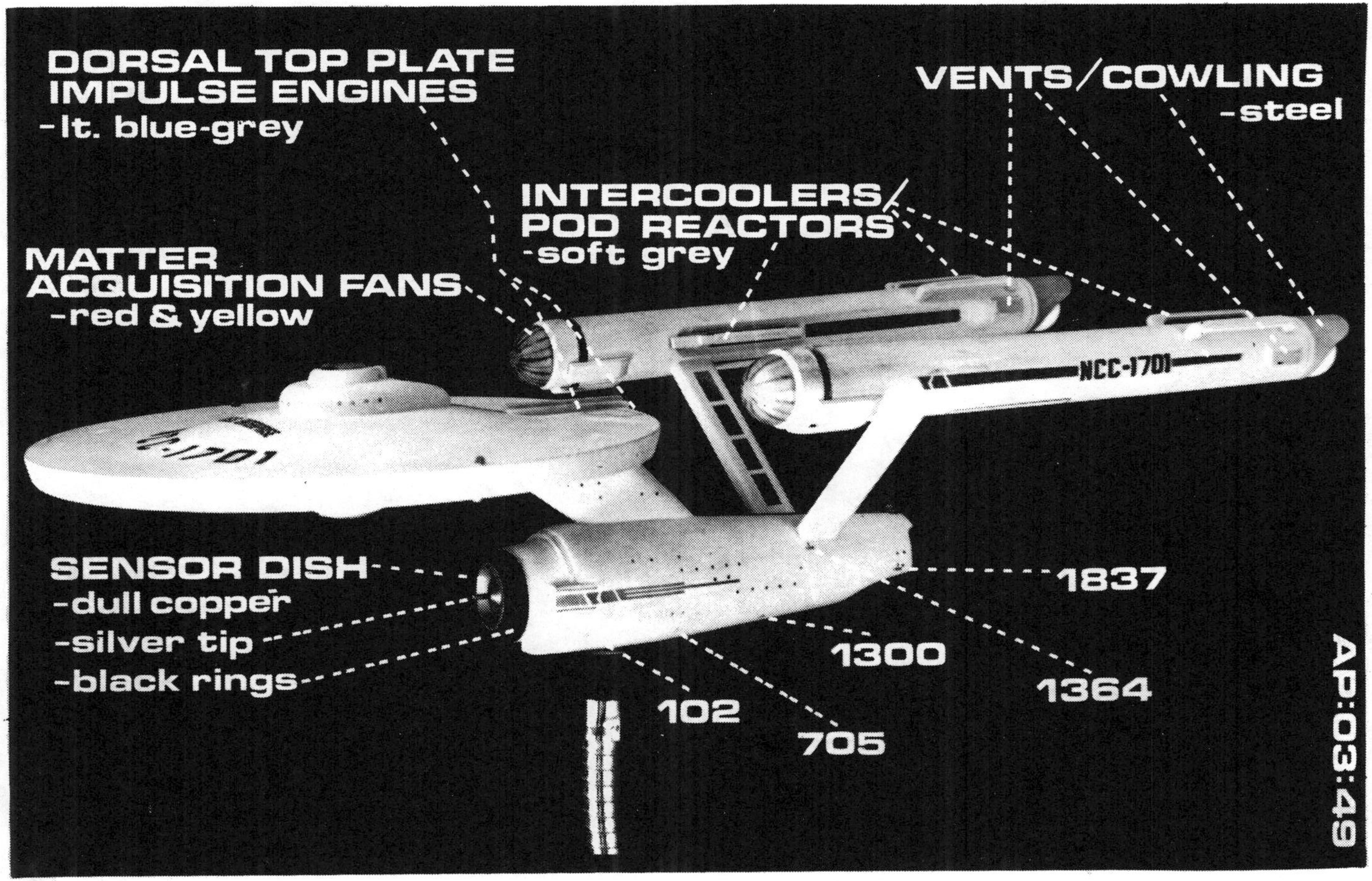
DORSAL TOP PLATE
IMPULSE ENGINES
-lt. blue-grey
VENTS/COWLING
-steel
INTERCOOLERS/
POD REACTORS
-soft grey
MATTER
ACQUISITION FANS
-red & yellow
NCC-1701
SENSOR DISH
-dull copper
-silver tip
-black rings
1837
1300
1364
102
705
AP:03:49

IN THIS BOOKLET:

- ASSEMBLY OF THE:

LS/100A
Digital Electronics System™

- INSTALLATION OF:
  Lighted Ports
  Sensor Arrays
  Running Lights
  Homing Beacon &
  WARP ENGINES
- SUPERDETAILING
- PAINTING & WEATHERING
- A FLEXIBLE BASE
- CENTERSPREAD ARTWORK

3

PAUL M. HEWITT / StarFleet Division

Corrections

SFAM ONE

AP:01:05
The 8 1/2” x 5 1/2” binder idea never caught on, and so it wasn’t produced.

AP:01:06
The “Quake of 2004”? Who knows. It could happen.

AP:01:08
List #2, note: brush/liquid glue is no more effective than tube glue (i.e. Testors). Decals are available by outside vendors. See *www.culttvman.com*

AP:01:09
For an update on the color of the Enterprise, see “What Color is the Enterprise?” in the Appendix section.

AP:01:10
See SFAM 3 and the Appendix for corrections that list proper colors and paint selections.

AP:01:11
A note to reinforce the importance of sanding the seams. Too many starship models today (at shows) still have exposed seams. Also--the importance of SAFETY using C.A. (Super, Crazy, glue).

AP:01:16
Added note: remember to wash all the plastic parts in soapy water to remove the casting mold release. This will help the paint adhere to the plastic.

AP:01:28
The actual angle of the nacelles (of the 11 foot filming miniature) is just under 90 degrees, but the AMT kit was designed for 90 degrees.

AP:01:30
Sand off the grid if striving for realism. Also reduce the sharpness of the edge of underside “ring” on the primary hull to make the underside as flat as possible.

AP:01:36
The “intercooler” grills were present during the series production. There were not present during the pilots.

AP:01:38
There are many off-the-shelf airbrushes on the market, and since 1977, they certainly cost more than $10-15.

AP:01:40
See the updated reference to the colors in the appendices.

AP:01:44
Rear, contoured area should be dark grey. Again, see the updated color information.

AP:01:42
There are many other masking materials that can be used today. Liquid mask, and Tamiya masking tape is highly recommended.

AP:01:44
Step # 24; the “balls” are actually the same color as the hull.

SFAM TWO

AP:02:10
The D7 Klingon Cruiser is actually 28 inches long overall. Not 29 inches.

AP:02:16 through AP:02:19
Instead of the “Flourette” bulb, there are many descriptions of LED light retrofitting on the *www.culttvman.com* website.

AP:02:20
Sorry, it’s not an antenna. It’s a disruptor.

AP:02:25
The little chrome "B.R. Tubes" were a detail added by AMT. They are not really on the 28" filming miniature.

AP:02:30 / AP:02:31
For the most part, this painting description is accurate. There were two variations of grey, but there was NO green coloring on this filming miniature.

AP:02:32
The Romulan Bird-Of Prey was a Wah Chang designed ship.

AP:02:34 through AP:02:37
Today, there are many small 5/8" diameter spheres, and half-spheres available that will negate the need to carve a piece of plex rod! Too labor intensive!

AP:02:40
As on page AP:02:16, etc., you can substitute a white or yellow LED. Also, the plasma weapon was omitted but can easily be scratchbuilt.

AP:02:42
The coloring of the Romulan BOP is of some uncertainty. Try various websites for newer information.

AP:02:46
When this book was done, "cupcakes", and other snack cakes came individually wrapped in very thin foil. Obviously, today, this has changed, but there are many other sources for this foil today. Remember to use caution when using a soldering iron, and to be aware of the fumes from melting plastic.

SFAM THREE

AP:03:04 through AP:03:13
SPECIAL NOTE: This section was retained for this second edition of this book. HOWEVER, this circuit board (LS-100-a) IS NOT AVAILABLE. The author and publisher decided to include this section for historical reasons, and to provide some modeling and electronic ideas--available only at that time (1978). Remember to be very cautious when using a soldering iron, as it can become VERY HOT! Use adult supervision if you are unsure, and/or inexperienced with these tools.

TODAY there are several circuit boards that replicate this first-ever use of electronics in a starship model, back in 1978. You can seek these unique electronics at *www. culttvman.com*

AP:03:16
Note: Isopropyl (rubbing) alcohol is a flammable substance. Use with caution.

AP:03:18
(#5) DO NOT remove the screens from the intercoolers. See page reference AP:01:36.
(#7) Plex rod is useful, and it does have an effective "glow" to it, but you may wish to seek out other spheres, and domes in 5/8" diameter from various sources.

AP:03:22
For any references to the LS-100a circuit board, see page update AP:03:04 (above). Note that filament light bulbs can be replaced with 3mm or 5mm (bright) white LEDs (white LEDs didn't exist in 1978, although we had wished for them at the time....).

AP:03:32
(#35) It is suggested that a better putty would be "Modeling Paste" made by Liquitex, and sold in most art stores. It is basically ground up marble, suspended in an acrylic base.

An excellent putty that you can thin with water, and actually brush on.

AP:03:36
The Estes (Rocket Co.) decals for their Enterprise flying rocket ARE NOT AVAILABLE. Do not write them asking for these decals. There are many other sources (prompted by this book) for decal sets for this kit, as well as it in other scales. Look for these at: *www.culttvman.com*

AP:03:38
Many of the markings indicated here are now included on decals sheets. See the resource section for information.

AP:03:38 through AP:03:41
Superdetailing: These diagrams were completed for this book by renowned master model maker David Merriman.

AP:03:40
A model car headlight lens, or model railroad "MV" lens may be added on the shuttle homing beacon for additional realism.

AP:03:40
The pylon connection to the nacelle of the 18" AMT kit is incorrect and would require major surgery to fix.

AP:03:40
There are two triangular panels at 10 and 2 o'clock in the underside of the primary hull. Consult your reference material. There is also a possible use of these mystery triangles mentioned in the appendix section.

AP:03:40
The sensor dish for the 18" AMT kit is inaccurate. Consult your references for creating a replacement.

AP:03:40
There was a subtle grid on the primary hull. It was drawn with pencil on the 11 foot miniature. When reduced in scale for the model kits it would be almost undetectable. Sand it off.

AP:03:42
(#1.C.) See the discussion about the hull color in the appendix of this book. There are excellent references of the Smithsonian Enterprise at The IDIC Page. See reference section.

ENTERPRISE DIAGRAMS ON AP:00:13 AND AP:00:14 BY ALLEN EVERHART.

About Allen Everhart

The Everhart Enterprise plans, until recently, stood as the definitive record of this fictitious craft. Ever since their publication I and others found utilized them as we set about scratch-building and detailing models.

Allen was a draftsman of the old school and spent endless hours stooped over his table with pen and pencil in hand. Allen was dedicated to his craft. He risked life and limb climbing to the top of a shaky ladder snapping the best possible shots of the studio miniature as it hung suspended high over a stair well in the Smithsonian Air and Space Museum so many years ago.

In the mid-80's, Allen joined the service in order to provide a better lifestyle for his family. His last job: working the below decks powder magazine of gun turret #2, aboard the Battle Ship Iowa. I never saw Allen squander a moment of his life with listless thought or inaction. Of those who knew and respected Allen Everhart, we try to meet the standard he established.

David D Merriman III

Resources, References, and Bibliography

Here is a collection of resources that could be useful to the readers of this book. Many of the books and magazines are out of print, but they are worthwhile to locate. It would be easy to expand upon this listing, but we have tried to limit the focus to the subject matter of this book.

BOOKS AND MAGAZINES

Edmonson, Harold A., editor. *Famous Spaceships of Fact and Fantasy.* Kalmbach Books, 1979. Second Edition, 1996, edited by Terru Spohn. Reprinted articles from *FineScale Modeler* magazine includes several chapters about building the Enterprise and Galileo. Also features an article about the studio miniature by Richard van Treuren.

Goldstein, Stan and Fred Goldstein. *Spaceflight Chronology.* Wallaby 1980. Now outdated, this book chronicled the history of spaceflight as envisioned when the first Trek feature came out. Illustrated by Rick Sternbach.

Irwin, Walter and G. B. Love, editors. *The Best of Trek.* Signet, 1978. Contains the text of Richard van Treuren's articles on Trek miniatures and models.

Jackson, Rick. *Spaceships at the Final Frontier.* Kalmbach Books, 2001. Modeler's guide to building many of the Trek ships.

Joseph, Franz. *Star Fleet Technical Manual.* Ballantine Books, 1975. Diagrams, registration numbers, and inspiration for new starship designs.

Joseph, Franz. *Star Trek Blueprints.* Ballantine Books, 1973.

Mandel, Geoffrey. *Enterprise Officer's Manual.* Intersteller Associates, 1980. Includes starship diagrams and technology concepts that may interest model builders.

McMaster, Michael. *Book of Klingon Plans.* 1976. Blueprints similar to Franz Joseph's starship blueprints.

McMaster, Michael. *Romulan "Bird of Prey" Blueprints.* Galactic Designs, 1976. More blueprints similar to Franz Joseph's starship blueprints.

Newitt, Paul M. *StarFleet Assembly Manual 4.* Starfleet Division, 1980. The guide to building the refit Enterprise.

Okuda, Michael, Denise Okuda and Debbie Mirek. *The Star Trek Encyclopedia.* Pocket Books, 1994. Revised 1997, 1999. Exhaustive reference to the various incarnations of Star Trek, this book features extensive information and illustrations of starships.

Star Trek Mechanics Official Guide 4. Star Trek Magazine, 1999. Published in Japan. Excellent, full color photo book featuring virtually every starship from the various series. Worth seeking out.

Trimble, Bjo. *Star Trek Concordance.* Ballantine Books, 1976. Revised edition, Citadel Trade, 1995. References to starship registration numbers and more.

Whitfield, Stephen E. and Gene Roddenberry. *The Making of Star Trek.* Ballantine Books, 1968. Features photos and diagrams of the Enterprise. Great read too!

MAGAZINES

Anderson, Howard A., Linnwood G. Dunn, and Joseph Westheimer. "Out-of-this-world Special Effects for 'Star Trek.'" *American Cinematographer,* October 1967. Behind the scenes on the original series with rare photos!

Asherman, Allan. "The Miniatures of Star Trek." *Star Trek Giant Poster Book.* Issue 11, July 1977. More information about filming miniatures.

Drexler, Doug. "The Smithsonian Report." *Star Trek Giant Poster Book.* Issue 10, June 1977. Excellent article about the Enterprise and Klingon studio miniatures at the Smithsonian with detailed photos. Highly recommended.

Fiebiger, Daniel. "Special Visual Effects." *Cinefantastique.* July, 1996. Rare photos and more information about the filming of the miniatures. Excellent!

McCullars, William. "Enterprise '64 The Real Builders of the Storied Starship." *Star Trek Communicator.* Issue 132, February, 2001. The definitive story of the Enterprise filming miniature. Must read!

McCullars, William. "Enterprise '64 Part 2 Building a Better Starship." *Star Trek Communicator.* Issue 133, June, 2001. More on the Enterprise as well as the K-7 Space Station. Must read!

McCullars, William. "Raiders of the Lost Starcruiser." *Starlog,* October 1989. Article about the Klingon filming miniature.

Merriman, Jr., David D. "Scratchbuilding the Starship U.S.S. Enterprise." *Scale Modeler,* April 1992. Excellent article about creating a lighted starship using vacuformed polystyrene and resin.

Sides, Roger. "1701 - The Ultimate Refit." *Sci-Fi & Fantasy Models #14,* September, 1996. Article documenting the 1991 restoration of the Enterprise studio miniature. Incredible photos.

Van Treuren, Richard G. "Star Trek Miniatures: The Starship Enterprise." *Trek #5,* 1975. An early article about the filming miniatures and model kits with rare photos. Reprinted in The Best of Trek without pictures.

Van Treuren, Richard G. "Star Trek Miniatures: The Other Spaceships." *Trek #6,* 1975. More about the filming miniatures and model kits with more rare photos. Reprinted in The Best of Trek without pictures.

OTHER RECOMMENDED READING:

Modeler's Resource Magazine. Published 4 times a year, this glossy magazine often features articles of interest to science fiction modelers.

Model Maniacs Magazine. Published once or twice a year in conjunction with Chiller Theatre conventions in New Jersey. Often features articles about spaceships and hardware.

Fine Scale Modeler. This magazine occasionally features science fiction subjects, but is mainly devoted to military subjects. Even so, most of the modeling articles feature information of use to all model builders.

PRODUCTS AND SUPPLIES

CultTVman's Hobbyshop
www.culttvman.net
A good selection of model kits, lighting products and decals.

Federation Models
www.federationmodels.com
(248) 544-4663 or (877) 688-8537
Largest online retailer of science fiction models. Large selection of starships, decals, and aftermarket parts. Highly recommended.

JTGraphics
157 Hampshire Rd.
Baltimore, MD 21221
(410) 574-3220
Decals for all your science fiction modeling needs. Custom decal printing available.

Polar Lights
www.polarlights.com
Currently producing Star Trek kits including the 1/1000 scale classic Enterprise and Klingon ships.

Fiber Optic Products Inc,
P.O. Box 779
Clearlake Oaks, Ca 95423
(707) 998-3903
www.fiberopticproducts.com
A good online source for your fiber optic needs.

Plastruct
1020 South Wallace Place
City of Industry, CA 91748
(626) 912-7016 or (800) 666-7015
www.plastruct.com/
Large selection of plastic supplies including fiber optic.

Aves Studios
P.O Box 344
River Falls, WI 54022
(800) 261-2837
www.avesstudio.com
Modeling putty and sculpting supplies.

Magic Sculp
3629 Voleyn Street
Carmichael, California 95608
www.magicsculp.com
Epoxy putty for modeling and sculpting

Testors
www.testors.com
Model paint, airbrushes, glue.

Micro Mark
(800) 225-1066 or (908) 464-2984
www.micromark.com
All the tools a modeler could ever want.

WEBSITE REFERENCES

The IDIC Page
http://members.aol.com/idicpage
William McCullar's comprehensive collection of studio model photos and information.

Model Reference Vault
www.cloudster.com
Phil Broad's exceptional archive of studio model photo references.

The Model Citizen
http://employees.csbsju.edu/rsorensen/modelcitizen
Roger Sorensen's collection of models, references and building information. The first sci-fi modeling site on the internet.

Ed Miarecki
www.edmiarecki.com
Photos of the restoration of the Enterprise studio model.

Starship Modeler
www.starshipmodeler.com
Articles and news for the science fiction model builder. Highly recommended.

CultTVman's Sci-Fi Modeling
www.culttvman.com
Articles and galleries of great science fiction models.

What is the REAL Color of The Classic Enterprise?

an update for the SFAM Special Edition

by Paul M. Newitt. April 2004

In an article by William McCullars in issue 133 of "Star Trek Communicator", Richard Datin states: "(Matt) Jefferies also furnished paint chips, for which I had lacquer-based paint custom-matched by a Fuller's Paint dealer on Olympic Blvd. in Los Angeles.....it was not a Ford or GM paint or primer as some 'learned' souls have decried. It was a flat finish—a light grey color with a light tint of green."

Through my association with well regarded Hollywood prop maker Richard C. Datin (the original builder of 1964 vintage 33 inch and 11 foot filming miniatures), I was able to decipher this original color. These are based on the only known remaining sample, which is an overspray "splatter" on the backside of one of Mr. Datin's original Enterprise drawings, from a touch up spraying of the 33 inch miniature. For modelers of the original Enterprise kits, this is a real find. To interpret this color is not easy. Upon a visit with Mr. Datin, we held the side of the blueprint with the overspray into the sunlight, and with commercial paint chips I brought, we were able to match the color, with an artist's eye compensating for the age and type of the paper, any fading and density variation of the paint. We then interpreted and confirmed this color against all color photos of the TOS ship at the Smithsonian (pre-restoration), and 1968 "Space Seed" EFX stills at the Anderson Company. This was a very serious and academic appraisal of this color.

Grey is a very elusive color, since it ABSORBS colors around it, such as bluescreen, or studio lights with colored gels,to create a "bias". Ektachrome-based 35mm film stock used in the original series had a blue bias—which the miniatures picked up from the bluespill of the bluescreen. If you look at the studio miniatures, and on TV, you'll notice a "falling off" of the bright, seemingly neutral grey on the nacelle tops—to colored, contrasty edges on the sides and undersides. This is the nature of grey tones that precipitate biased inherent color, especially varied tones that are darker than you think, even when brightly lit. This is a trick to fool you into believing that miniatures are actually "real". Matt Jefferies was a master of color and tonal variation for sets and miniatures.

Since it would be easy for most modelers to obtain a good reference, I have provided a several "house paint" swatches to match to your model paints. This is for matching purposes only, and there is no intent to use actual house paint.

The following WalMart samples ("ColorPlace" paint) have been prioritized by Richard Datin:

1 choice—"Concrete", set # 632, # 96321
2 choice—"Armory Gray", set # 631, #96311

Note: "Concrete" is an olive drab based color, including a light neutral, cold grey. "Armory Grey" is close, but has the slight influence of a warmer grey. For those not able to go to a local WalMart we've have provided you with a sample on the back cover of this book.

Other "house paint" swatches close to the "Concrete" swatch include:

Behr #450E-3, Southern Breeze (needs a bit of cold gray to match)
Glidden #50GY-55/066 Distant Haze (needs a bit of neutral gray)
Sears Weatherbeater #KK192 Silver Sage (is a bit too green)

Several model paints either match or are close to this color, including Testor's Model Master, Tamiya Acrylic, and PollyScale. These paint listings are meant for mixing and airbrushing. Remember that a 10-20% variation from the Walmart samples will not be readily noticed on the 18-22" classic Enterprise models, since grey, by nature has a wider latitude than most solid colors. The 18-22" Classic Enterprise model kits are about 1/5th of the 11 foot miniature, and 1/2 size of the 3' miniature, both which were painted the same grey-green color. If advanced modelers wish to be very precise, you can add about 1-3% more white to compensate for scale, but it will be unnoticeable to the naked eye.

Tamiya Acrylic paints:

To match "concrete", use XF-11, "J.N. Green" (this is the closest match of all model paints). To match "Armory gray", use XF-20, "Medium Grey" (lightened about 20%).

For the Nacelle grilles, and dark detailing, use XF20 Medium Grey

---

PollyScale paints:

F414113, "Reefer white"
414 332, "New Gravel Grey" (or you can use "SP Lettering Grey")
414209, "Burlington Green"

Mix as follows: 1 part green; 6 parts grey; 1-2 parts white

Or to match "concrete" use "I.J.A. Lt. Gray, # 505274 (but a little dark), or 505326 Pale Blue Gray (close).

Note: be careful adding white with water-based paints, since the addition of white has a dramatic influence on a base color; much more so than using Testor's. PollyScale is primarily a model railroad paint. Floquil makes equivalent colors in their "non-water-based" paint as well. Floquil paints were used extensively on the miniatures in ST:TMP (except the Enterprise), for their fine-grain, realistic application.

Testor's Model Master paints:

FS 17875, Insignia White, Testors #1745
FS 34159, Sac Bomber Green, Testors #1793
FS 36440, Flat Gull Grey, Testors #1730
For Grilles use: FS 36081 Euro I Gray

Mix as follows (to match "Concrete"): 1 part green; 6 parts grey; 23-25 parts white.

If you use spray cans instead of an airbrush, you can use first use (regular) Testors 1224 Green as a base, then 1233 Flat Aircraft Gray overspray on top until you get the amount of gray/green to match the paint swatch. Remember to vary the gray while spraying to give that "tonal variation" look. Allow at least a day for the green to dry first.

The idea here is to first get a real muddy looking military-looking olive drab, then to lighten it up with white to the desired shade. This color "concrete" has a definite green look to it, but it also has a powdery cold grey quality to it as well. Be careful when adding the white, because 20 parts white is not enough, and 30 is way too much, and washes out the color. Experimental samples were done on white styrene sheets, then tested on a neutral grey piece of plastruct, to simulate any possible bias by a darker undercoat. When properly applied, Testor's Model Master covers very well, and there was no real difference in color on a medium grey surface.

When you are completed with the painting of your ship, to weather it, use model railroad chalks in black or dark gray, streaking the chalk from front to rear, then seal with semi-gloss or flat clearcoat. The other option is to airbrush Floquil "grime" diluted about 30%, and hold the airbrush at least 12 inches from the surface, with a front to rear motion.

Good Luck on painting and detailing your Classic Enterprise!!

The Studio Models

The Enterprise today at the National Air and Space Museum in Washington D.C. Photo courtesy of Mike Warshaw.

THE ENTERPRISE

• The Enterprise filming miniatures were constructed in late 1964 for the original pilot by Richard Datin and his crew, based on designs by Matt Jefferies. The models were reworked in late 1965 for the second pilot adding internal lighting to the 11 foot ship. The distinctive nacelle lighting and other details were added in mid-1966 for the regular production of the series.

• The series often combined effects footage from the pilots and regular production resulting in an inconsistent look for the starship.

• The 11 foot miniature was constructed using vacuformed plastic sheets on plywood ribs for the primary hull, wood planks for the secondary hull, and sheet metal on a plywood frame for the nacelles.

• The large studio model is 134 inches long. The saucer is 60 inches in diameter. The nacelles are just over 72 inches long. The secondary hull is 49 inches long, not including the sensor dish.

• There is a grid on the primary hull of the eleven foot miniature. It was drawn on with pencil and is virtually invisible except upon close inspection. The grid and additional weathering were added after production on the series started.

• The smaller 33 inch miniature was a fixture in Gene Roddenberry's office in the 1970's. It was reportedly loaned out for a convention appearance and never returned.

• The 11 foot miniature was donated to the Smithsonian Air and Space Museum in 1974. By this time, the original lighting no longer functioned and was replaced with an inferior system. The nacelle domes and sensor dish had been lost and was replaced at this time.

• In preparation for a major Trek exhibition, Ed Miarecki oversaw the restoration of the 11 foot model in late 1991. The model was completely unassembled and rebuilt. The original lighting effects in the nacelles and other internal lighting were restored. The top of the primary saucer was left largely untouched but the rest of the model received a new paint job. The heavy weathering is not considered accurate to the original miniature.

• The 11 foot Enterprise is currently on display in the gift shop of the National Air and Space Museum in Washington DC

THE KLINGON BATTLECRUISER

• Two similar Klingon miniatures were built by AMT based on designs by Matt Jefferies. One was used for the series effects footage. The second miniature was is believed to have been used by AMT to tool their model kit.

• The Klingon filming miniature was donated to the Smithsonian in 1974 along with the 11 foot Enterprise. It has only been displayed during the 1992 Trek exhibition.

• The Klingon Battlecruiser measures 28.5 inches long and at the widest point is 19.5 inches.

• During preparation of the Star Trek 2 series in 1978, Paramount borrowed the original Klingon miniature from the Smithsonian. Though it was not used in production, the ship was repainted with a new color scheme before being returned to the Smithsonian. For the 1991 exhibition, Ed Miarecki painted the ship grey in attempt to return the model to its original look.

• The second Klingon miniature was given to Stephen Whitfield Poe by Gene Roddenberry. It has since been auctioned and changed hands several times. This model is believed to have the original, undisturbed paint scheme.

THE ROMULAN BIRD OF PREY

• The filming miniature was built by Wah Chang for the first season episode "The Balance of Terror" and the footage was reused in a later episode. Little is known about this model as it disappeared soon after production but it is believed to have been approximately 3 feet across. The nacelles were also illuminated.

For more about these miniatures visit William McCullar's website, The IDIC Page at *http://members.aol.com/IDICPage*

Tips for the Polar Lights Kits

Modeling the Polar Lights Enterprise
tips by Jay Chladek

The Polar Lights U.S.S. Enterprise model kit provides parts and decals necessary to do several variations of the ship. The instructions aren't completely clear on what is needed to build each variant and this list will help sort out those details.

PILOT ENTERPRISE FROM "THE CAGE"

The first version of the ship is featured in the pilot episode, and some footage was occasionally used during the series.

• Fill the recessed panel lines for the triangle shaped panels on the bottom of the saucer and remove the raised steps, as the pilot ship didn't have them. Remove the grooved pattern from the rectangular areas where the arrowhead decals on the secondary hull go. Also fill in the recessed line on the dorsal connector.

• Use parts 16, 17, 20 and 24 for the pilot ship warp engines. Use parts 35 and 23 for the spiked warp nacelle endcaps. Part 37; the hydrogen collector detail pieces that go on the inside aren't really needed since the hydrogen collectors were not clear on the pilot ship anyway.

• For the secondary hull, use the big deflector dish (part 8) and the ungrooved hangar bay doors (part 10).

• For the saucer, use the tall bridge dome piece (part 28) and install the lower sensor dome without the nipple (part 30) into the lower saucer. Install the saucer light pieces (part 32) as normal. But when the saucer is together, sand the clear bulbs flush with the top and bottom of the saucer and keep the molded in bulbs on the saucer instead. Remove the grooved detailing from the impulse engine (part 3) as well.

• The rear endcaps on the warp engines of both the pilot models were the same gray shade as the rest of the ship. The bussard collectors were painted a dark reddish copper on the real ship and there was no swirling light effect. The spikes were gold in color. The color accents on the bottom of the warp nacelles may still be there, but this isn't certain.

• The decal instructions for the kit are pretty good, so follow their recommendations for which markings you need to add. Be aware that the font used for the lettering and numbers of this ship are different from what was on the series model, so be careful not to swap decals.

• The model wasn't lighted, so use only the darkened window decals if you want to make your model representative of what was on the studio model.

PILOT ENTERPRISE FROM "WHERE NO MAN HAS GONE BEFORE"

Throughout the entire series run, this is the version of the ship that was seen in the opening credit sequence. Pilot ship shots were also seen in most every episode as well alongside shots of the production version of the ship. To build this version of the ship, follow the guidelines for the first pilot version except for the following changes:

• Use the pilot two version warp nacelle endcaps (part 25). Again, part 37 is considered optional and not really needed since the bussard collectors were solid again.

• The spikes on the collectors were silver in color on this ship.

• The model had some lights for the windows, so you can use either the darkened window decals from the pilot model or some of the series style lighted windows. The rest of the decal instructions are accurate.

PRODUCTION ENTERPRISE FROM THE TV SERIES

This was the model seen in the vast majority of episodes. The kit represents pretty much all of the details found on this version of the ship very well and will build up into a nice replica.

• Use parts 18, 19, 21, 26, and 27 for the engines and parts 36 for the non-spiked endcaps. Part 34 is intended for the swirling lighted detail on the inside.

• For the secondary hull, use the small deflector dish (part 9) and the grooved hangar bay (part 11).

• Remove the raised navigation lights on the gray saucer parts. DO NOT sand off the navigation lights from the clear parts. Use clear part 31 for the lower saucer sensor dome with the nipple on it and the short bridge dome (part 31).

• Coloring is the same as the other ships for the most part, but the production ship was accented and weathered slightly. For the darker patches on the central strut and the bottom of the warp engines, mix a color slightly darker then the gray base shade (add a few drops of darker gray to stain it slightly). If you desire to add a grid to the model, carefully add one with a very sharp pencil using a compass and straight edges as guides. These panels all appear to have been drawn onto the studio model, not etched into the surface.

• The rear endcaps on the series ship were dark gray.

• The triangle patches on the bottom of the secondary hull should be a medium gray color. Also, use a pencil or a fine line black pen to outline them in black for improved appearance.

THE "MIRROR UNIVERSE" SHIP

• The markings provided in the kit for the ISS Enterprise are nice, but the studio model was never marked like that. Still, the unique markings look cool. The main shot of the ship during the flip-flop sequence is of the second pilot version, featuring the spiked nacelle endcaps and the taller bridge dome. But, all the remaining distance shots of the ship in orbit are of the series version. So, it's entirely up to you as to how best to detail this particular version of the ship. A nice compromise might be to do the model with the spiked endcaps, and tall bridge dome, while adding the series version parts for the rest of the model. The modular nature makes such a compromise easy and nobody can say the model is wrong anyway.

GENERAL TIPS

• Even though this model is a snap kit, glue all the parts. Putty and sand the seams as you would any other kit. Trim the alignment pins short to allow easier assembly. Attaching the nacelles to the pylons can be tricky so be careful. If you choose to light the model, you will need to cut a grove into the pylons to run your wires.

Modeling the Polar Lights Klingon D-7
tips by Steve Iverson

• The Klingon D-7 Battlecruiser from Polar Lights is not quite as accurate as their Enterprise. You should check reference photos such as those online at The IDIC Page or The Model Builder's Reference Vault to help you make your corrections.

• The initial run of this kit had a problem where the "neck" of the ship joins the body causing it to lean to one side. Carefully trimming the part will correct this.

• The "platform" of the bridge section is too tall, especially the rear section. The vertical domed cylinder is also too tall.

You should be able to cut these parts up and carefully rebuild them to match your reference images.

• The two angled “radiators” that attach to the shuttle bay are incorrectly mounted at a 45 degree angle. The angle should be closer to 75 degrees, or almost vertical to the shuttle bay.

• The engine nacelles should be angled inward about one quarter inch. This could be tricky to achieve so be careful.

• The smooth and ribbed inserts for the leading edges of the hull fit flush to the hull. These should be recessed.

• Finally, the main hull of ship needs to be thickened. This could be achieved by layering sheet styrene on the top and bottom of the ship. Again, consult your references.

• The model is intended to be 1/1000 scale like the Enterprise, but depending on what references you use, the model may actually be about 10% too large.

The Romulan Bird of Prey

• Playing Mantis has announced that a plastic Romulan Bird of Prey will be released as part of a Johnny Lighting diecast collection. This will reportedly be 1/1000 scale like the Enterprise and D-7.

• An excellent 1/1000 scale Bird of Prey is available as a resin garage kit from Starcraft Models. Check Federation Models for availability.

The K-7 Space Station Kit

BASIC ASSEMBLY

The K-7 Space Station is a relatively easy kit to build. First of all, refer to the back cover of the paperback "The Trouble With Tribbles" by David Gerrold. There's a color photo of the K-7 and you'll notice that there are no "ribs" on each round "pod section." You can eliminate these by cutting or sanding, then apply pie shaped pieces of .040 thick styrene to the tops and bottoms of the pods. Otherwise, the assembly is not difficult, but will require some body putty and much sanding.

HOMING BEACONS

Here is a schematic of the flasher for the four homing beacons located at the top of each cone. By using this system with four yellow sub-miniature light emitting diodes, you can achieve a flashing rate of 3 seconds "on" and one second "off," exactly as in the episode. The tops of each cone can be drilled out for each LED and the center beacon will have to be supported by a copper rod of 3/16" diameter. Then place a clear bead on top of the LED to for a spherical beacon.

PARTS LIST
- 555 Timer I.C.
- 10 Mf capacitor
- 150 K resistor
- 75 k resistor
- 4 - 550 ohm resistors
- 4 - LED's yellow or white
- Diode (1N4001)
- 9V transistor battery or battery eliminator

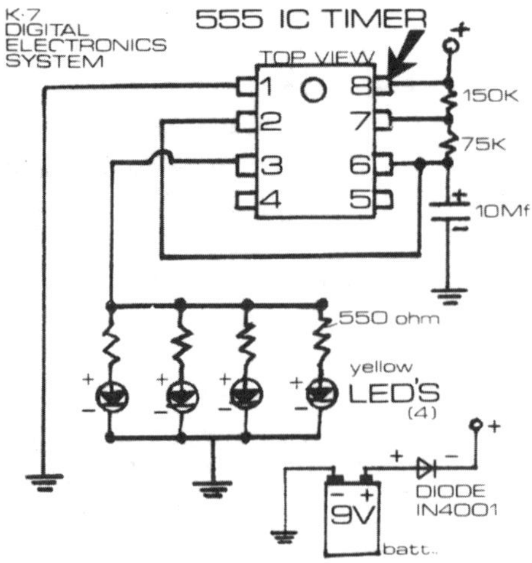

LIGHTED PORTS

Located on each "pod" as well as the central section, is a cone with many lighted ports. This is achieved by drilling out ports at random on the cone with a small drill and a #65 bit. Use the inside of the rectangles as a guide for each hole, then insert 40 mil fiber optic into each hole allowing about 1/4" to stick out that will be trimmed off with nail cutters after the cones have been painted. The lenses at the end of each pod can be lighted with fiber optic also. The other end of the fiber optic can be run down each "tube" to the center to a light source, or any other method you choose. Make sure the ends of the fiber optic are cut perfectly flat, and that they face the light source directly. Bulbs for the light source should be 9-12 V grain of wheat size or white LEDs.

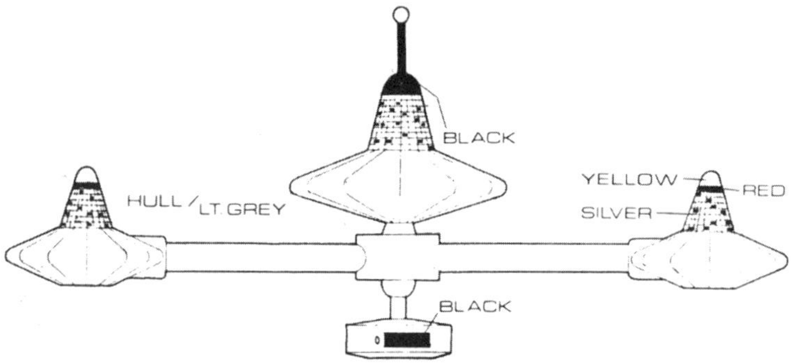

PAINTING

The side elevation of the K-7 (drawn by Geoffrey Mandel) shows the color schematic. The hull should be painted a color matching the photo of the back of Mr. Gerrold's book. This is a combination of concrete color, with silver and some grimy "weathering" from space dust, debris, etc. First, spray (preferably airbrush) it a grey/concrete color, like Floquil "Platinum Mist"; finally, a thinned out coat of Floquil "Grime." This will dry to a flat finish, so spray with some photo-matte spray for a semi-matte surface. You might try some "tonal variation" if using an airbrush. That is, make variations in the intensity of the colors. (Remember, paint all of the parts a flat black to keep any light from coming through the plastic first.)

Building the NOMAD prop as seen on "ENTERPRISE"

For some associates of mine who work on "Enterprise", I decided to make a prop/model that would fit into the pre-TOS timeline scenario. I chose the NOMAD satellite that was launched from Earth in August of 2002, since this fit perfectly into the show's pre-TOS timeframe, and it would make a subtle connection between both series for the sharp-eyed fan.

I first spent a lot of time studying the NOMAD diagram as seen in the original series episode "The Changeling" (originally aired on 9/29/67). I also studied NASA websites that showed commercial and telecommunication satellites from that era. These satellites were silver-chrome body colors with gold crinkled foil and mechanical pieces painted either yellow, red or white. I did take the liberty of choosing a grey color for the body, as "if" the TOS art department in 1966 were to actually make one of these, it would probably be a consistent grey color used on the show.

I chose to use a 1 inch PVC tube as the main body. I placed a few smaller tubes in one end to form the taper that forms the "nose" (acquisition sensor), and liberally puttied it

with liquitex "modeling paste". The actual nose is a "Uniball Vision" pen end cap. The ring (" sperographic analysis comp.") and its struts were made from stock styrene. The main tube runs through the body, and towards the rear to the engine ("primary power system"). I left that extended, to cut down later.

The irregular flat strips on the main body ("data assimilation") were styrene strip I-beams with the front ends rounded off. The tubes with the spheres at the ends were actual aluminum tubes, about 3/16" diameter, and bent gently around a curved table edge, to not buckle them. I then drilled holes through the previously attached styrene strips, and slid them in at an angle, to match that of the diagram. Once these were dry, I glued black beads to the tube ends.

The antennas and solar panels ("selective amplifier screen") were made by carefully cementing two types of scribed styrene together. First, I took a grid pattern sheet, and rolled it with a piece of pipe, to give it a "memory" to stay curved. I then took a parallel strip type sheet of styrene, and also rolled it. I took both rolled pieces and glued one into the other—both at the same curvature. These were oversize pieces, so I could cut them out later, like a piece of half-pipe, to the exact size I needed.

The flat "canoe paddles" ("tracking screens") were made from a .060" piece of styrene, and carefully sanded with sanding sticks to a thin outer edge. I then ran a small rod through each end to guarantee a solid support for these paddles, and then added the trim detail that look like arrows.

The rear engine ("primary power system A/N 57") is actually a cap from a spray pump bottle that you find everywhere. I had a large half-sphere that fit this cap, and then puttied it all together, with the rod in the center of it. The rear, which is not in the drawing, was detailed to look as though a stream of ions would emanate, so I used the end piece of a red Uniball Vision pen as the thruster.

The paddles and curved panels were covered with "bare metal foil" in both gold and aluminum. This foil is intended for 1/24th model cars, but I've used it for everything but cars! The spheres were also covered in gold foil, with lots of crinkle, to match that of photos of other satellites I saw on the JPL

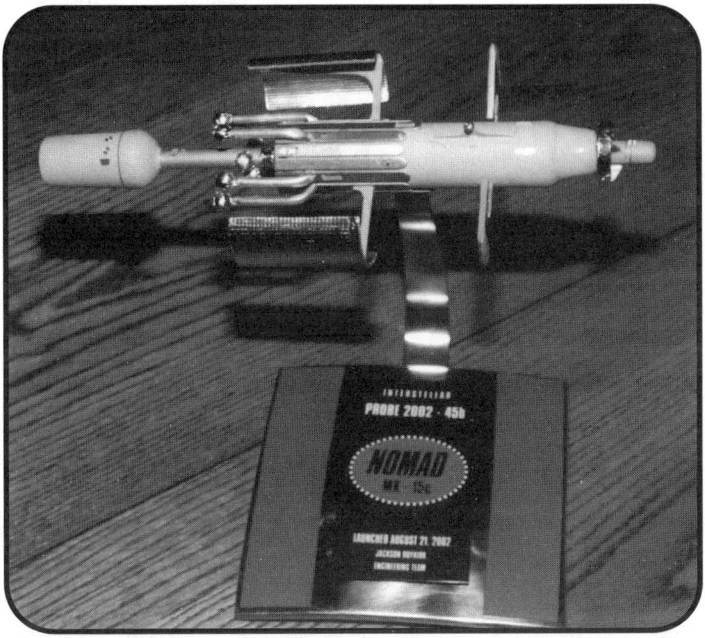

website. In the back area, where the rod extends through, I packed more gold foil, as if to protect some sensitive instruments. The body is hand painted with pollyscale grey paint, and the yellow is also hand painted. The rear of the ion engine is painted a metallic dark aqua, to give it "that mysterious high-tech look of 2002".

At the hobby shop, I chose some photo-etched detailing for a model car. All the small details are actually brake and gas pedals, and gas caps in 1/24th scale! The struts that hold the large curved panels are thick styrene (.080"), cut to the same curved shape as in the diagram. I wanted to follow the diagram to the letter, although some detailing, and angles leave room for some leeway. I felt that this model was successful in its replication, since the reaction at the art department was quite good!!

The base is simple. The bottom is an aluminum candle base found at Target, turned upside down. The vertical support is actually an aluminum spoon that I bent at a gentle, but steep angle. The support was "J.B. Welded" to the base underside. A slot was cut into the NOMAD, and it sat just perfectly on it. The paper graphic I made was not included when the show was filmed.

The use of this prop was a snap decision, as it sat on a desk in the art department for some time. I'm not sure what prompted its use, but the strange part about it was the episode "Dead Stop" (which premiered on Oct. 9, 2002) for which it was filmed—was shot in August of 2002! Needless to say, the opening scene with my model in semi-closeup was very fulfilling, and well worth this effort.

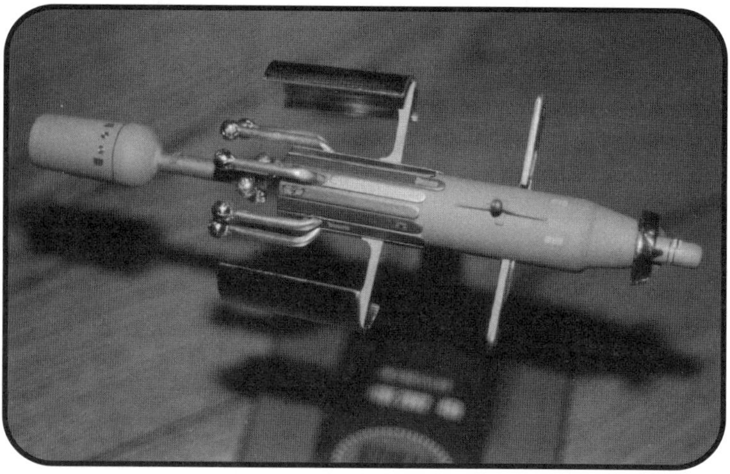

Constitution Class Shuttlebay How To

In 1982, I was asked by a small group making a student SF film to make some miniature sets. One of these miniature sets was a 1/35 scale shuttlebay set for the AMT Galileo shuttlecraft kit.

This project involved some interpretation "in three dimensions" to decipher two dimensional drawings into a model. Using color slides of pages from the Star Fleet Technical Manual, I projected the side elevation of the bay on a sheet of posterboard. I set up the Galileo model on a table for a size reference and traced various details onto poster board.

From that point, I finished the rear wall and clamshell doors. The rear wall was made to be high enough to attach the upper observation deck "pods" and hallways. After the doorways were cut out, doors were made from extra cardboard stock.

The observation deck, and the rearward "observation control towers" that are on the upper deck, were all made at once. Tracings from the Tech Manual were used to make eight sets of side window sections, and four fronts, all with windows cut out. The "observation gallery" required making a

bulkhead about every 4-5 inches for a consistent, solid shape. The insides of the gallery and control tower pods were covered with frosted mylar and small incandescent model railroad lights were used to illuminate them.

Finally, I had to make the ceiling with backlit "skylight". Since this section had to be removable, I had to engineer this model to be self-supportive in two parts. This was difficult, because a 3-D version of this had to be extrapolated only by a trial and error fit of curved posterboard. The frosted mylar ceiling piece was fit in place and Plastruct metal rods covered in white styrene were used as the support "ribs" for the mylar. Three small box-shaped ceiling pods finished this area off. To support the whole top structure, I had to make a simple frame from moulding, and attach wires to even out the height of the ceiling section.

A BBQ motor made the turntable rotate, and the landing pattern was painted red and yellow. The entire interior was painted with Floquil "primer" grey by hand. The cardboard is a little unforgiving to paint like this—since it reveals texture to the surface, however, it does give it that "Trek-Set" look. When a floodlight is held over the frosted mylar of the ceiling, it produces very effective lighting, as shown in the photos.

This was a fun project, and I recommend making sets from any film or TV show in scale as a new avenue to explore in SF/ST modeling.

Model Concepts

The following diagrams are reprinted by an exclusive agreement from the authors, Geoffrey Mandel and Doug Drexler. These diagrams are originally from the "Star Fleet Handbook", and the "USS Enterprise Officer's Manual", published in January, 1980 by Interstellar Associates. We include these diagrams for the inspiration of modeling and detailing opportunities.

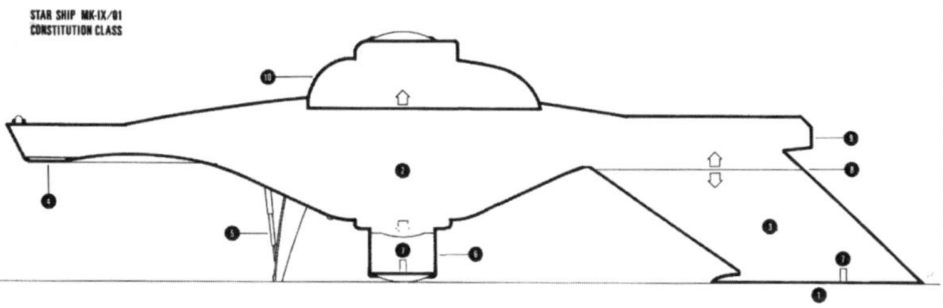

EMERGENCY LANDING PROCEDURE

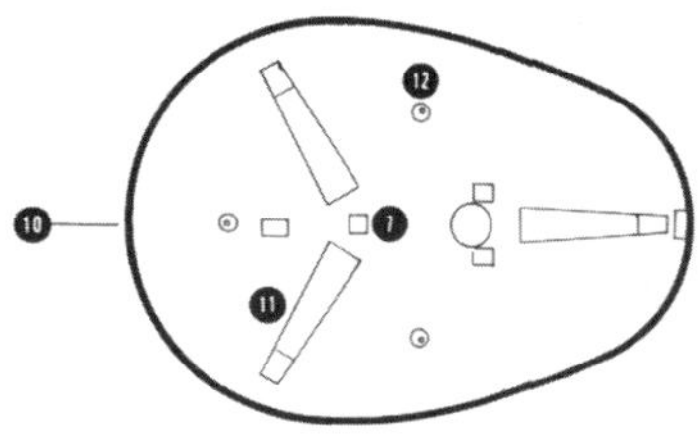

BOTTOM VIEW

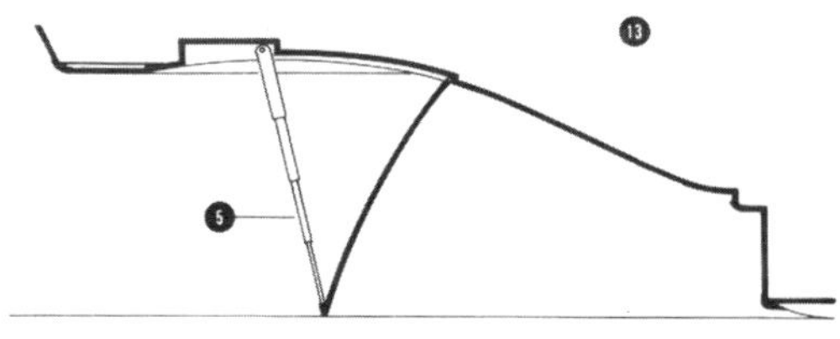

EMERGENCY LANDING PROCEDURE

This is a depiction of “cracking out of there with the main section”, as envisioned by Geoffrey Mandel. This concept pre-dates the hull separation idea for ST:TMP, and would make for an interesting Constitution Class Primary Hull landing scene diorama, with a dorsal landing trail in the landscape. The bridge section can separate from the primary hull as a lifeboat. One possibility for those “mysterious triangles” on the saucer underside, is the use as landing legs, somewhat like a C-57D.

FLIGHT DECK CONTROL TOWER

This close-up of the rear of the Constitution Class hangar area, provides detailed information and speculation for the functions of these features. Note the use of the red rectangle on the underside as a landing target, and the launch tubes for probes. This would make an excellent 1/35th scale close-up shuttlebay model in the same scale as the AMT Galileo model kit.

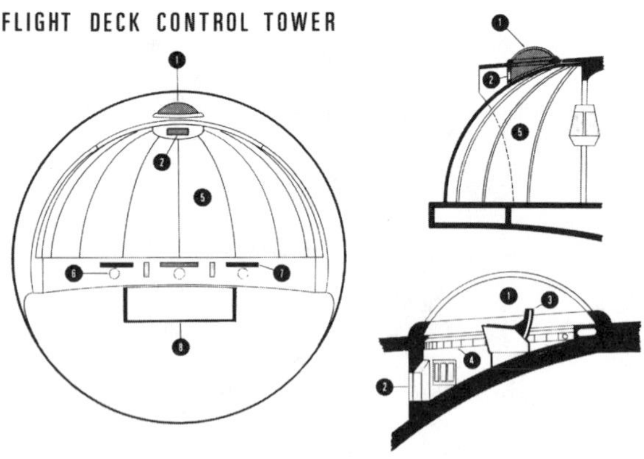

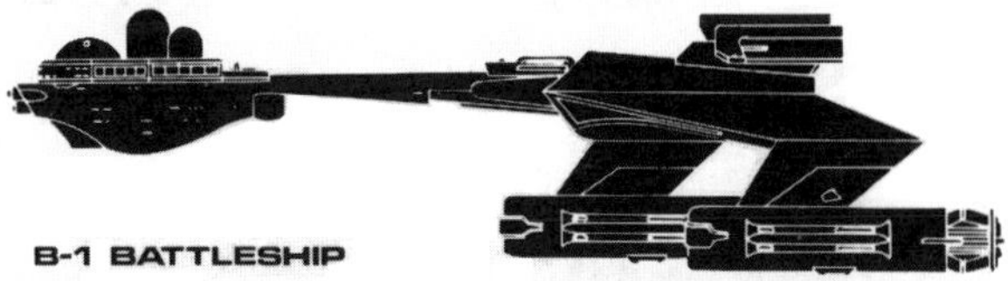

B-1 BATTLESHIP

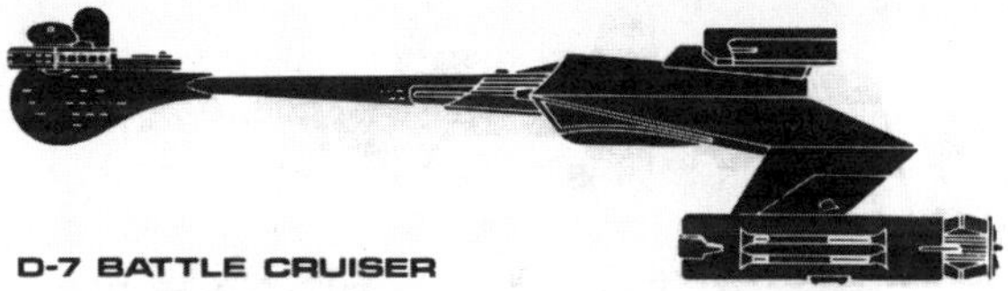

D-7 BATTLE CRUISER
NOTE: K'T'INGA CLASS IS SIMILAR

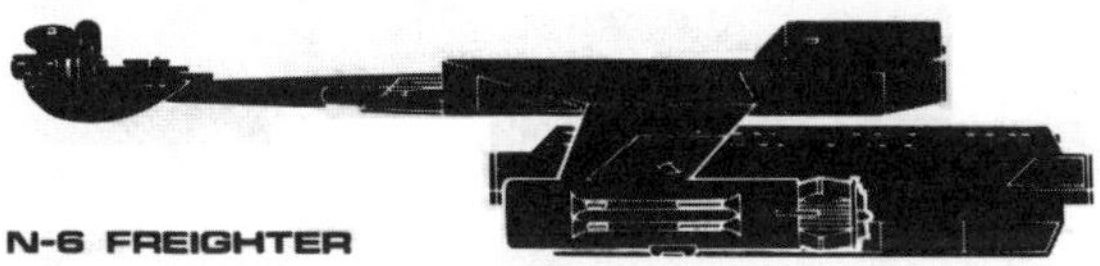

N-6 FREIGHTER

WARSHIP RECOGNITION

For those into Klingon ships, these speculative variants of the D-7 class are potential candidates for kitbashing. The "B-1 Battleship" is the Klingon answer to Star Fleet's triple-nacelle ships; with a third Graf Unit (nacelle) mounted in the center of the hull underside. Note the modified battlehead, too. The N-6 Freighter is a cargo carrier, with the use of an underslung "K'ontainer". This would involve the use of the D-7 kit's nacelles, and struts, with a kitbashed hull section. The neck would remain the same, with a rebuilt head, with the top-side bridge section retained.

About Paul M. Newitt or "What Happened after the SFAM series?" A personal experience.

The original limited editions of the SFAM series were published in the late 1970's, and were distributed across the country. As a result of this exposure, I was invited to a Trek convention in New York (ST: America) in 1978. I visited with Geoffrey Mandel (author of the "Medical Reference Text", and both ST Star Maps); and Doug Drexler, as well as others into "tech-trek" on the east coast.

An Enterprise A finished and lit, built with the use of SFAM 4.

In 1980 I'd just graduated from college at UCD School of Design, and was off to live in Southern California. These books opened a lot of doors and caught the attention of many effects professionals in Hollywood, including Greg Jein, Andrew Probert, Rick Sternbach, Mark Stetson, and others that worked on Star Trek: The Motion Picture. I was fortunate to have visited several TV and movie sets (including TRON, AirWolf, and Trek IV), and learn how scale models, props, and film sets were designed and built in Hollywood shops.

Living in San Diego led to several visits with Franz Joseph, the creator of the "Star Fleet Technical Manual". SF/ST fans living in one house sponsored many working weekends, including scale models and Japanimation costuming and props for conventions we attended throughout Southern California. I also did graphics work with "Fantasy Cars" in San Diego, including some detailing of a one of the Barris Batmobiles, the "Spinner" and "Deckard's Police Car" from Bladerunner, and the "Leslie Special" from "The Great Race". A "secret" side trip in 1984 to the Arizona desert included seeing the outdoor sets for "Jedi" ("Blue Harvest"), and chatting with the stars.

In 1982, I built this shuttlebay in 1/35th scale, based on plans in the SFTM, for an amateur film project. It is made from artboard, and frosted mylar for the skylight. Each control room is also lit.

Published in 1984, StarFleet Assembly Manual FOUR was a major project that began in San Diego, and took over a year to produce. I had the support and advice of the ship's designer, Andrew Probert, to help in the authenticity of modeling this ship and making corrections on the production model kit. This was a leap in this modeling genre, and included the engineering and production of a circuit board using CMOS technology, and a “sequencing randomizer” for the warp nacelles. This 70 page book included a red binder, and was the first time Trek plastic “retrofit” parts were offered to modelers. This was an extremely popular book, and was so popular, it was compromised by the unfortunate fact of piracy, being copied many times over. It was this book that generated the new branch of scale modeling in the Science Fiction arena, which has spurred many outside companies that make parts, decals and electronics.

Here I am getting to know the new Moonliner just before it was lifted onto its permanent platform. Tomorrowland in Southern California, on "press day", May 22, 1998. Guest of Imagineer David Mumford.

StarFleet Assembly Manual 5, a little known two-part article in the HJS Publication "Enterprise" (issues #1, April 1984, and #2, June 1984), described the assembly and modification of the "Avenger Class" ship. Since the AMT Reliant kit superceded this modification of the Enterprise kit, the series of articles was not continued. Also in 1982, I scratch built a classic shuttle bay, in scale with the 1/35th AMT Galileo Shuttlecraft kit, out of cardboard and wood framing for an amateur film. Other projects included making props and models from current TV shows and movies for our costuming groups in S.T.A.R. San Diego.

This is the Nautilus weathervane that I made in memory of WDI Imagineer and 20K expert Tom Scherman. The plaque reads: "In Memory of Tom Scherman", the reverse plaque reads: "Guiding the Winds to Discovery Bay". The bottom cups do spin. This weathervane is six feet tall. This weathervane now resides at the exit que to the 20K attraction at Tokyo DisneySeas® in Japan.

In the late 1980's, my Trek scale modeling experience took a hiatus while I worked on professional exhibit design projects and other interests. The mid 1990's included opportunities to make a few TV show props, including the Stearman Biplane desk model for Harm's office in "JAG", a "Rocketeer" diorama for the producer, and a NOMAD satellite for Mayweather's cabin on "Enterprise". These latter opportunities were most fortunate through my good friends Geoffrey Mandel and Doug Drexler, both "former New Yorkers" and award-winning artists that have worked on Trek productions and publications for many years. At that time I also became interested in Disney's "Moonliner", and I built several for Disney Imagineers. Among them was Imagineer Tom Scherman, who had his North Hollywood apartment done up like the Nautilus. He was the ultimate "Vernian", and the unparalleled expert on Disney's "20,000 Leagues Under The Sea". To honor the loss of Tom to cancer in the late 1990's, I built an elaborate 6 foot tall Nautilus weather vane, which today resides at the exit cue of the 20,000 Leagues attraction at Tokyo DisneySeas.

I'm standing in the H.G. Wells room at "Travels Through Time", based on the conservatory from the George Pal movie "The Time Machine". This version of the Time Machine was reproduced by Don and Mary Coleman, and is perfect in every detail.

This is the starship bridge at "Travels Through Time" Science Fiction Museum. It was loosely based on fan blueprints, and includes a captain's chair, flashing lights on panels, and backlit screens.

1999 brought on a huge project in Old Sacramento called 'Travels Through Time" This was a 3200 square foot science fiction museum that I was asked to design and oversee construction of. The museum was open until 2002 and was featured in the center section of the March 2000 issue of STARLOG magazine. This project included a Jules Verne room, a replica Time Machine built by Don and Mary Coleman, a crashed martian war machine, an alien planet scene with famous built up resin models from 1950s movies, and a full size section of the classic bridge with captain's chair, famous for its appearances at many Los Angeles conventions. This museum educated many young people, and provided lots of awe and enjoyment for many visitors.

My next project involved the fine folks at the "Admiral Collection"; a high-end collectibles store at "Star Trek: The Experience" at the Las Vegas Hilton. I produced two limited edition ship series for them. Eleven of the Masterpiece Replicas Enterprise "D" starships were retrofitted by myself (and my engineering team) to provide remote control lighting, synchronized with sounds on six channels. The second edition was a limited set of the 28" long D-7 Klingon ships, the same size as the original filming miniature from TOS, carefully reproduced with a new base design. Both ships were very popular with serious collectors, and sold quickly.

This is the 28 inch brass Enterprise "D", by Masterpiece Replicas. Eleven of these ships were retrofitted by myself with remote control lights and sound effects for the "Star Trek: The Experience" in Las Vegas. All sold in less than one year, at $3,000 each.

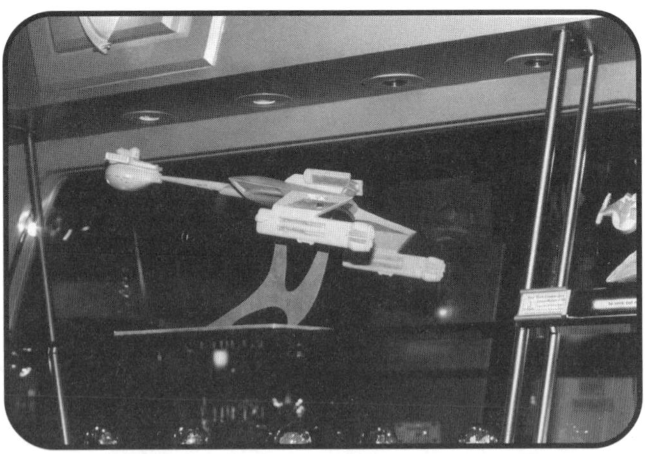

One of six D-7 Klingon ships reproduced from a Greg Jein model. Great care with detail was taken to accurately reproduce these ships for "Star Trek: The Experience". The support base is based on a Bat'lith Klingon weapon, and was designed by Paul M. Newitt.

I'm holding the Vor'cha ship is one made by Greg Jein from the original filming miniature molds, made of fiberglass, and it is over 30" in length. The brass Enterprise is 28" in length, and is 1/1000th scale, the same scale as the Polar Lights TOS-E kit

Over the years, I've had the opportunity of meeting, talking to, and knowing several fans and enthusiasts involved with the making of SF and ST scale models. Many enduring friendships have evolved from these books - from artists in Hollywood, engineers at NASA, and to everyday hobbyists that enjoy learning about how to make their kits more exciting. The driving theme to doing these books was to bring life to otherwise "boring" scale models. To enhance the Trek personal experience with an animated starship only makes that world a little more real. As a designer first, and fan and modeler second, I am not only proud from producing those books, but to have had the many opportunities that have resulted from them. Two of my proudest moments include a flight from Seattle in 1984 with Gene Roddenberry who complimented my copy of SFAM 4 and to learn that all of the SFAM series books are currently on file at the Star Trek art department at Paramount.

For more of Paul M. Newitt's work, visit
www.culttvman.com/sfam

Acknowledgments

I would like to thank and acknowledge the following people for their assistance on the original SFAM series (1977-1978):

A.J. Desmaretz for his printing advice, Al Sirois for his wonderfully visionary artwork, Mark Finlay for the StarFleet Division Yards and Klingon Scout Ship artwork, Judy Montgomery (Bryer) for her fan inspiration, Frank Kurson for his complete support and enthusiasm, David Merriman for his professional direction and artwork of Enterprise superdetailing, Geoffrey Mandel for bringing this all to a higher level, and to Doug Drexler for his undying focus on original Trek. Special thanks to Leonard Suligowski for his faith in my work. I also wish to thank my family for their support, and for the original typing of the SFAM 1-3 series. Special thanks (in memory of) Franz Joseph Schnaubelt for his guidance with the SFTM, and personal friendship. The original LS-100a engineering was developed by Terry Whittier, which, at that time was a senior lab tech at IBM (LS-100 prototyping completed by Ken Montgomery). I also wish to thank all the fans at that time that sent letters of support, advice, and reaction to all the enjoyment that these booklets provided. Special thanks to Bjo Trimble for her support hard work to save Trek (and those wonderful “Equicon” fan-run conventions); and “Allison” at S.T. Welcommittee; S.T.A.R. Sacramento, and Jean (Harshman) of S.T.A.R. San Diego.

Since the original publication of these books in 1977-78, other words of thanks go to:

Steve “Cult TV Man” Iverson for his vision and complete support of this new publication; Andrew Probert & Rick Sternbach for his design mentoring, Mark Stetson and Greg Jein for their scale model advice; Phillip Lublin for his friendship and resin-casting expertise; David Saul of the “Admiral

Collection" (ST:TE) in Las Vegas; Don & Mary Coleman for their "Time Travel"; Gary Kerr for his never-ending quest for autocadding the "E"; and to the original Enterprise model-maker Richard C. Datin for his gracious assistance in obtaining the "E" color.

This book is dedicated to the design genius and memory of Walter "Matt" Jefferies.

Paul M. Newitt
9 April, 2004

The publisher would like to thank Paul Newitt for all his inspiration over the years. Without the original StarFleet Assembly Manual, there may never have been a CultTVman. Additional thanks to David Merriman and Anthony Taylor for pushing and prodding me to produce products for model builders, William McCullars for his enthusiasm, knowledge, and sharing his photo collection on the web, Jay Chladek for help with model wisdom and building, and Dave Metzner and the folks at Polar Lights for bringing back Trek models. Finally thanks to all who contribute to the CultTVman website, the Hobbytalk forum, and entire internet modeling community.

Steve Iverson
www.culttvman.com

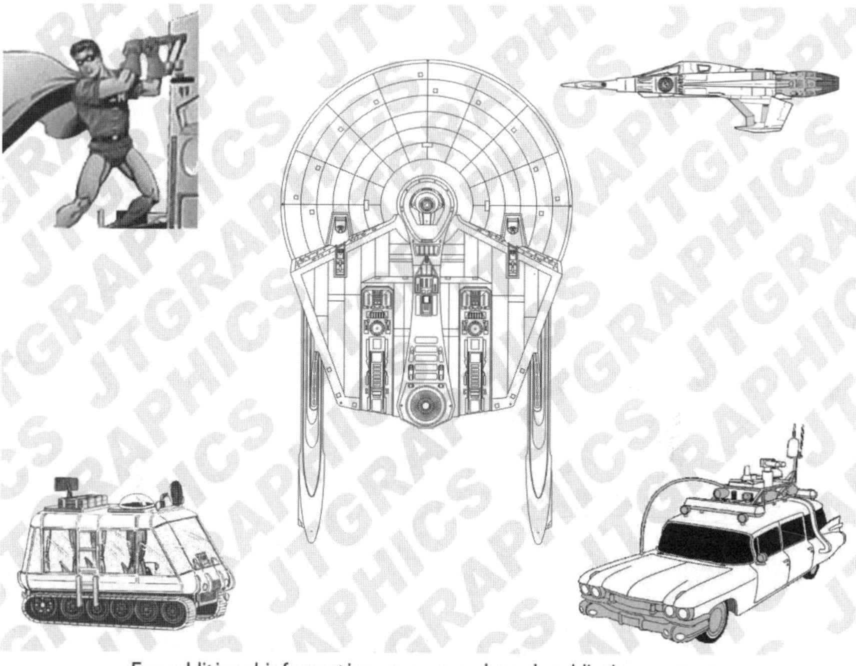